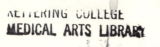

DIAGNOSTIC ULTRASOUND

Physics, Biology, and Instrumentation

DIAGNOSTIC ULTRASOUND

Physics, Biology, and Instrumentation

Stewart C. Bushong, Sc.D., FACR, FACMP

Professor, Department of Radiology
Baylor College of Medicine
Houston, Texas

Benjamin R. Archer, Ph.D., DABR, DABMP

Associate Professor, Department of Radiology
Baylor College of Medicine
Houston, Texas

With 194 illustrations

Mosby
Year Book

St. Louis Baltimore Boston Chicago London Philadelphia Sydney Toronto

Publisher: David T. Culverwell
Developmental Editor: Christi Mangold
Assistant Editor: Mary R. Taylor
Editorial Assistant: Cecilia F. Reilly
Editorial Project Manager: Jolynn Gower
Production Assistant: Pete Hausler
Book Design: David Zielinski

Printed in the United States of America

Mosby-Year Book, Inc.
11830 Westline Industrial Drive,
St. Louis, Missouri 63146

Library of Congress Cataloging-in-Publication Data
Bushong, Stewart C.
    Diagnostic ultrasound : physics, biology, and instrumentation/
  Stewart C. Bushong, Benjamin R. Archer.
      p.   cm.
    Includes index.
    ISBN 0-8016-0394-3
    1. Ultrasonic waves. 2. Ultrasonics in biology. 3. Ultrasonic
  equipment.   I. Archer, Benjamin R. II. Title.
  QC244.B87   1991
  620.2′8--dc20

                                                        90-25538
                                                          CIP

Stewart C. Bushong dedicates this book to several dear friends who left a considerable legacy of love and influence.

Edmund Burke
Herman Bushong, Jr.
Nell Clark
Merrill Collier
David Hodges
Walter Horan
Robert Kelly
Ralph Kidwell
Peter Kuijpers
Bruce Lunsford
Robert MacIntyre

Benjamin R. Archer dedicates this book to the memory of Vera Inez Archer and to Christopher Lockett Archer, world class swimmer and son.

Preface

Ultrasound was introduced into diagnostic imaging in the early 1960s. However, it took approximately 20 years for it to develop into a principal diagnostic imaging modality. Its inherent advantages—relatively low cost, non-invasive and easily conducted examinations—have led to an ever increasing demand for ultrasound technology. Today diagnostic ultrasound stands as the first line of diagnostic imaging for many patients.

This rapid growth rate is attested to by the increasing sales of diagnostic ultrasound equipment and the increasing number of sonographers. The first examination administered by the American Registry of Diagnostic Medical Sonographers was in 1975. By 1983, a total of 4,349 sonographers had been registered. In 1990, 5,748 people sat for the certification examination. Currently, there are over 16,000 registered sonographers.

We have been teaching the physics of diagnostic ultrasonography for 15 years. Our students are sonographers in training at the Houston Community College and radiology residents at Baylor College of Medicine. This textbook is an outgrowth of the teaching activities, lecture notes, and handout material that we have developed over these years. Although this text is primarily intended for diagnostic ultrasound students and radiology residents, students from a variety of specialties that use medical ultrasound should also find it beneficial. Among those that are specifically served by this text are sonographers and physicians in cardiology, obstetrics, and surgery. Medical physicists and ultrasound engineers should also find this text useful.

Throughout, our primary goal has been to make the physical concepts relevant to ultrasound image production and interpretation. It is our conviction that anyone professionally involved with ultrasound imaging must understand, not simply memorize, the physical principles involved in the creation of the ultrasound image.

To make these physical concepts more identifiable, important definitions, symbols, and units are set off from the text with a screen. Some of the physical concepts are unavoidably expressed as mathematical formulas. To help eliminate barriers to understanding these equations, numerous worked examples are incorporated throughout the text. Chapter 2 is intended to provide the necessary mathematical background for any student to begin the successful study of ultrasound physics.

In assembling material for this text, we are deeply grateful to a number of sonographers and educators who have shared their suggestions and knowledge with us and in many cases have provided valuable reviews of the many drafts of this text: Jim Bindon, Houston, TX; Jana Craig, Houston, TX; Terry Ellis, Bothell, WA; Debra Gaden-Robertson, RDMS, Houston, TX; Joe Hamilton, Houston, TX; Denise Hazzlerigg, RDMS, Houston, TX; Lisa Joseph, RDMS, Houston, TX; Robert Malone, M.D., Houston, TX: R.J. Matthews, Houston, TX; Betty Napo, Houston, TX; Jonathan Ophir, Ph.D., Houston, TX; Sherri Quast, Tustin, CA; Cheryl Ross, RT, RDMS, Vancouver, WA; Cynthia Smith, Dallas, TX; Lydia Tamez, Houston, TX; Ralph Winters, Tustin, CA; Diana Zientek, R.T., Tomball, TX.

We are also deeply grateful for the care and attention that our medical illustrators, Spencer Phippen, Carol Larson, and Leon Kairewich paid to every detail of every figure. Finally, our sincere appreciation to Judy Matteau Faldyn, not only for her timely and efficient processing and reprocessing and reprocessing of the manuscript, but also the pleasant manner in which she completed it by fending us off from time to time.

The physics of diagnostic ultrasound can be mastered. As with any other endeavor, you will gain knowledge and understanding based on your input time and effort. As you use this text, please feel free to communicate with us about any sections that should be changed, simplified or expanded. To our fellow educators, we know you are with us in our attempt, through this volume to MAKE PHYSICS FUN.

Stewart C. Bushong, Sc.D.
Benjamin R. Archer, Ph.D.

Contents

1 Introduction to Ultrasound, 1

The Nature of Ultrasound, 1
Historical Introduction, 1
    Industrial Applications, 2
    Medical Applications, 2
Review Questions, 4

2 Fundamentals of Physics, 5

Mathematical Background, 5
    Fractions, 5
    Decimals, 5
    Significant Figures, 5
    Algebra, 6
    Scientific Notation, 7
    Rules for Exponents, 8
    Binary Number System, 8
    Graphing, 10
Units of Measurement, 10
    Standard of Length, 12
    Standard of Mass, 12
    Standard of Time, 13
    Systems of Units, 13
Scientific Prefixes, 13
Review Questions, 14

3 Basic Physical Quantities, 15

Velocity, 15
Acceleration, 15
Newton's Laws of Motion, 15
    First Law, 15
    Second Law, 15
    Third Law, 16
Work, 16
Power, 17
Energy, 17
    Mechanical Energy, 18
    Potential Energy, 18
    Chemical Energy, 18
    Electrical Energy, 18
    Thermal Energy, 18

Nuclear Energy, 19
Heat, 19
Mechanical Equivalent of Heat, 20
Review Questions, 20

4 Fundamentals of Waves, 21

Electromagnetic Waves, 21
Mechanical Waves, 21
    Transverse Waves, 21
    Longitudinal Waves, 24
    Surface Waves, 24
Wave Properties, 24
    Wavelength, 24
    Frequency, 25
    Period, 26
    Velocity, 26
    Wave Equation, 26
    Density and Compressibility, 27
        Density, 27
        Compressibility, 27
    Phase of a Wave, 28
    Huygen's Principle, 31
Review Questions, 31

5 Ultrasound Power and Intensity, 33

Power, 33
Intensity, 33
    Amplitude, 33
    The Decibel, 35
Intensity Specification, 36
    Spatial Variation, 37
    Temporal Variation, 37
Review Questions, 39

6 Interaction of Ultrasound with Matter, 40

Attenuation, 40
Attenuation Interactions, 42
    Absorption, 42
        Viscosity, 43
        Relaxation Time, 43

Frequency, 43
Refraction, 43
Scattering, 44
Diffraction, 44
Interference, 45
Reflection, 45
Angle of Incidence, 45
Acoustic Impedence, 47
Illustrative Example, 49
Review Questions, 50

7 The Ultrasound Transducer, 51

Piezoelectric Effect, 51
The Single Element Transducer, 51
The Case, 53
The Crystal, 53
Damping Material, 54
Matching Layer, 54
Quality Factor, 56
Review Questions, 58

8 The Ultrasound Beam, 59

Characteristics of Ultrasound Pulses, 59
Pulse Repetition Frequency, 59
Pulse Repetition Period, 59
Pulse Duration, 59
Duty Factor, 61
Spatial Pulse Length, 61
The Wave Front, 61
Near Field and Far Field, 62
Focusing the Ultrasound Beam, 65
Review Questions, 68

9 Ultrasound Resolution, 69

Axial Resolution, 70
Spatial Pulse Length, 70
Ultrasound Frequency, 70
Damping Factor, 73
Computing Axial Resolution, 73
Lateral Resolution, 73
Transducer Size, 73
Beam Width, 75
Frequency, 75
Computing Lateral Resolution, 76
Review Questions, 76

10 Pulse-Echo Imaging, 78

Range Equation, 78
Amplitude Mode, 79
Brightness Mode, 82
Motion Mode, 87
Review Questions, 87

11 Real-Time Transducers, 88

Characteristics of Real-Time Images, 88
Line Density, 88
Frame Rate, 90
Pulse Repetition Frequency, 90
Maximum Depth of Image, 91
Sector Angle, 91
Interdependence of these Characteristics, 91
Mechanical Real-Time Transducers, 92
Electronic Real-Time Transducer Arrays, 93
Linear Array, 93
Phased Array, 97
Annular Array, 97
Specialty Ultrasound Probes, 97
Focusing Real-Time Ultrasound, 98
Electronic Focusing, 99
Dynamic Electronic Focusing, 102
Steering the Ultrasound Beam, 102
Review Questions, 104

12 Image Processing and Display, 105

Pulser, 105
Receiver, 105
Amplification, 105
Near Gain, 107
Delay Control, 107
TGC Slope, 107
Far Gain, 107
Enhancement, 107
Reject, 107
Signal Processing, 107
Demodulation, 107
Enveloping, 108
Scan Conversion, 108
Analog Scan Converter, 108
Digital Scan Converter, 110
Preprocessing and Postprocessing, 113
Display Devices, 113
Image Recording, 116
Polaroid Images, 116
Multiformat Cameras and Laser Printers, 116
Other Recording Modalities, 117
Review Questions, 117

13 Doppler Ultrasound, 119

Doppler Mathematics, 119
The Doppler Equation, 119
The Doppler Angle, 121
Doppler Modes, 123
Continuous Wave Doppler, 123
Pulsed Doppler, 124
Aliasing, 127

Spectral Analysis, 128
Duplex Doppler, 130
Color Flow Devices, 130
Review Questions, 131

14 Image Analysis, 133

Reverberation, 133
Shadowing, 134
Reflective Shadows, 135
Attenuation Shadows, 137
Edge Shadows, 137
Enhancement, 137
Displacement, 137
Misregistration Artifacts, 139
Multipath Artifacts, 139
Beam-Width Artifacts, 140
Distortion, 141
Aliasing, 142
Review Questions, 142

15 Quality Control, 143

Quality Control Test Devices, 143
AIUM Test Object, 143
Group A, 144
Group B, 144
Group C, 144
Group D, 144
Group E, 144
SUAR Test Object, 144
Sensitivity, 145
Uniformity, 145
Axial Resolution, 145
RMI Tissue Phantoms, 146
Routine Measurements and Observations, 148
Sensitivity, 148

Distance Calibration, 148
Caliper Calibration, 149
Registration, 149
Dead Zone, 149
Axial Resolution, 150
Lateral Resolution, 150
Time-Gain Compensation (TGC), 151
Display Characteristics, 151
Visual Inspection, 151
Program Frequency, 152
Review Questions, 153

16 Biologic Effects of Ultrasound, 155

Mechanism of Action, 155
Thermal Effects, 155
Mechanical Effects, 156
Effects on Simple Structures, 157
Molecular Responses, 157
Water, 157
Macromolecules, 158
Chromosomes, 158
Cells, 158
Observation on Animals, 158
Observation on Humans, 159
Therapy Patients, 159
Physical Therapy, 159
Hyperthermia, 159
Surgery, 159
Dentistry, 159
Diagnostic Patients, 159
Dose-Response Relationships, 160
Clinical Safety, 160
Review Questions, 162

Appendix, 163

1 Introduction to Ultrasound

THE NATURE OF ULTRASOUND

Ultrasound is a form of energy consisting of mechanically produced waves with frequencies above the range of human hearing. This is a technical definition that most likely contains many unfamiliar terms. As the text proceeds, each of these will be explained in detail. But as a philosopher once said, "A journey of 1000 miles must begin with the first step." So, let's take that first step.

Everyone has heard the shrill (high frequency) sounds of a police whistle above the noise of traffic (low frequency) that he or she directs. The whistle produces mechanical waves (vibrating air molecules) that can be detected by the human ear/brain system. Amazingly, this system responds to vibrations on the order of 0.00000000001 meters, less than the radius of an atom. These vibrations, transmitted through the air, are audible or sound waves because they have frequencies between 20 and 20,000 Hertz, within the range of human hearing. As we will learn in Chapter 4, Hertz (abbreviated Hz) is the unit of frequency and indicates the number of wave cycles that pass a point each second. However, every ear does not respond exactly the same to high frequencies. People over 50 years of age can seldom hear frequencies in excess of 15,000 Hz, and for some the cutoff is below 10,000 Hz.

At least part of the traffic noise cannot be heard by anyone, regardless of age. It is below the threshold of hearing and can only be detected by vibrations that are felt. Thus frequencies below 20 Hz are termed **subsonic** or **infrasound** which means below sound.

Have you ever tried in vain to hear a dog whistle (Fig. 1-1) while your favorite pet howls and becomes as hyperactive as a teenager at a rock concert? The frequencies emitted by this whistle, invented in 1876 by Sir Francis Galton, are beyond the human hearing range but audible for a dog. These frequencies, above 20,000 Hz, are called **ultrasound.** The dog whistle was the first ultrasound instrument.

Bats use ultrasound for inflight guidance and for tracking flying insects. They do this by emitting ultrasonic waves and listening for echoes from various reflecting objects. Using ultrasound, bats can fly around navigational barriers and converge on their next meal.

Diagnostic ultrasound uses frequencies far higher than those emitted by the dog whistle or those within the range of human hearing. Frequencies ranging from one million Hz to ten million Hz are commonly employed in diagnostic ultrasound. Table 1-1 summarizes the different categories of sound by their respective frequencies.

Some of you may have studied x-ray physics. X-rays are wavelike phenomena. Ultrasound is also a wavelike phenomena. However, as Chapter 3 will show, ultrasonic waves are fundamentally different from the electromagnetic waves that constitute x-rays, light, and other forms of electromagnetic radiation. It is interesting to note that Wilhelm Conrad Roentgen, who discovered x-rays, also worked with ultrasound earlier in his career.

HISTORICAL INTRODUCTION

Less than 1 year after their discovery in 1895, x-rays had become widely used in diagnosis and therapy throughout the world. The evolution of equipment for the production of x-rays and imaging applications was also extremely rapid. By contrast, the development of ultrasound technology proceeded at a snail's pace. The Curie brothers, Jacques and Pierre, first discovered the piezoelectric effect that produces ultrasound in 1880. This effect is simple to describe. When certain crystal-like materials are subjected to a mechanical pressure they produce an electrical pulse. The reverse piezoelectric effect was demonstrated only a year later by the Curies. They showed that when an alternating voltage is applied to a piezoelectric crystal, the crystal vibrates and produces high frequency pressure waves (ultrasound). Thus, the fundamentals necessary to create

Fig. 1-1 The dog whistle was the first man-made instrument for ultrasound.

Table 1-1 Various categories of sound by frequency range

Name	Frequency range
Infrasound	Below 20 Hz
Audible sound	20-20,000 Hz
Ultrasound	Above 20,000 Hz
Diagnostic ultrasound	1,000,000-20,000,000 Hz

the modern ultrasound transducer that can send and receive ultrasound were known before the discovery of x-rays. However, diagnostic ultrasound was not used clinically until the early 1960s.

Industrial applications

One of the first practical uses of ultrasound occurred in the unsuccessful attempt at locating the *Titanic* which sank in 1912. In 1916, Langevin developed a method of underwater communication and researched methods for locating submarines using ultrasound. Ultrasonics improved during the Second World War with the military development of SONAR (SOund Navigation And Ranging). Fig. 1-2 illustrates the use of sonar for detection of submarines. A beam of ultrasound is transmitted from the surface vessel into the ocean. If the ultrasound intersects a submerged object, a small portion of the beam will be reflected back to the surface vessel where it is detected and analyzed. The distance to the object is calculated by noting the time required for the ultrasound to travel to the reflecting object and return.

The development of sophisticated sonar equipment during World War II helped spur the growth and development of medical ultrasound. During this same time, Firestone in the United States and Sproule in England developed pulse-echo ultrasound equipment for flaw detection in metals.

Medical applications

The first practical ultrasound imaging unit was designed by K.T. Dussik of Austria in 1937. He used two transducers positioned on opposite sides of the head to measure ultrasound transmission profiles in hopes of visualizing the cerebral ventricles. However this method proved impractical because of attenuation by the skull.

In 1949 Douglass Howry and W.R. Bliss, engineers at the University of Denver, constructed the first B-mode scanner from surplus naval sonar equipment, an oscilloscope, and a quartz transducer. As the transducer moved in a straight line above the skin's surface, echoes were displayed as white dots on the oscilloscope screen. The dots formed a crude picture of the underlying anatomy. The initial sonograms were made with the transducer and the object examined submerged in water in a laundry tub. Later, a metal stock tank was substituted for the laundry tub, and the quartz transducer was replaced

Fig. 1-2 The use of ultrasound to detect submarines resulted in SONAR.

by a transducer of lithium sulphatemonohydrate. Another worker in this laboratory, Posakony, evaluated transducer materials and replaced the lithium compound with others such as barium titanate and lead zirconate titanate. The latter material is almost exclusively used in transducers today.

By 1954, Howry and Holmes constructed the third B-mode scanner called a sonoscope from a B-29 gun turret. Patients climbed into the water-filled turret with lead weights hanging from their waists to prevent them from floating. The transducer carriage rotated 360° around the patient while a submerged 2.5 inch diameter transducer moved laterally across the carriage. These were the first compound scans produced as a combination of linear and circular movements of a transducer. Images were displayed on an oscilloscope with a persistent phosphor. The degree of fading of the images over time indicated the variations of echo intensity. This approach was

probably the first attempt to display ultrasound images in a gray scale format.

In the late 1960s, the water tank and gun turret were replaced by a semicircular pan with the bottom removed and a plastic membrane inserted in its place. The patient was pressed against the membrane, and mineral oil was used as a coupling agent between the patient and the membrane. The transducer was then moved in a semicircular path through a water bath on the opposite side of the membrane. With this device, reasonably good ultrasound images were obtained for a variety of abdominal conditions including the detection of cysts and abscesses in the liver.

For his efforts, Howry is recognized as one of the pioneers of diagnostic ultrasound. John Wild, another pioneer, demonstrated in 1951 that ultrasound could distinguish between normal and malignant tissues. He also used modified military surplus equipment. The first contact static B-mode scanner commercially manufactured in the United States was produced in the early 1960s by Physionics.

Thus, it took more than half a century for ultrasound to develop from SONAR to commercial medical applications. But once the potential and advantages of ultrasound were recognized, developments became more rapid and impressive. Even more exciting prospects lie ahead.

Review Questions: Chapter 1

1. Define or otherwise identify the following:
   a. Infrasound
   b. Ultrasound
   c. Audible sound
   d. Electromagnetic wave
   e. Piezoelectric effect
   f. SONAR
   g. Transducer

2. Explain how ultrasound differs from infrasound and audible sound.

3. Draw diagrams to compare the use of SONAR in finding submarines and the use of diagnostic ultrasound in finding kidney stones.

2 Fundamentals of Physics

MATHEMATICAL BACKGROUND

Any study of basic science, such as physics, must be accompanied by a basic knowledge of mathematics. Physics uses mathematical terms and equations to describe natural phenomena. Mathematics provides exactness and certainty that can never result from subjective descriptions of an event or interaction. For example, if a man drops a stone (Fig. 2-1) from the top of a tower, one can measure the time it takes the stone to fall and then compute the height of the tower. This calculation is made possible by a basic law of physics called the Law of Gravity. In mathematical terms the describing equation can be written as:

$$h = 1/2\ gt^2 \qquad\qquad (2\text{-}1)$$

where h is the height of the tower in meters, g is the local acceleration of gravity (9.8 meters per second²) and t is the measured time of fall in seconds

Note that g and t are known, and t is squared (t × t) in this equation. Similar laws that mathematically describe the fundamental nature of ultrasound will be explained. But first a basic math review is recommended.

To be a successful sonographer, the student should review and understand each type of problem presented in this review.

Fractions

x/y: x is called numerator; y is called denominator.

Addition and subtraction require a common denominator.

$$\frac{3}{5} + \frac{1}{6} = \frac{18}{30} + \frac{5}{30} = \frac{23}{30}$$

$$\frac{2}{3} - \frac{1}{2} = \frac{4}{6} - \frac{3}{6} = \frac{1}{6}$$

Multiplication: multiply numerators together, then multiply denominators.

$$\frac{4}{5} \times \frac{7}{3} = \frac{28}{15} = 1\frac{13}{15}$$

Division: Invert the second term and then proceed as in multiplication.

$$\frac{3}{8} \div \frac{2}{5} = \frac{3}{8} \times \frac{5}{2} = \frac{15}{16}$$

$$\frac{4}{5} \div 10 = \frac{4}{5} \times \frac{1}{10} = \frac{4}{50} = \frac{2}{25}$$

Decimals

Fractions in which the denominator is a power of 10 may be readily converted to decimals:

$$7/10 = .7 \qquad 46/100 = .46$$
$$7/100 = .07 \qquad 271/10000 = .0271$$
$$7/1000 = .007 \qquad 5431/10000 = .5431$$

If the denominator is not a power of 10, the decimal equivalent can be found by division.

$$\frac{3}{8} = 8\overline{)3.00}^{\ .375}$$

$$\begin{array}{r} .375 \\ 8\,)\,\overline{3.00} \\ \underline{2\ 4} \\ 60 \\ \underline{56} \\ 40 \\ \underline{40} \end{array}$$

Significant figures

In addition and subtraction, round to the same number of digits as the entry with the least number of decimal places.

Example:
$$\begin{array}{r} 55.1324 \\ 6.2 \\ \underline{8.07} \\ 69.4024 \end{array}$$

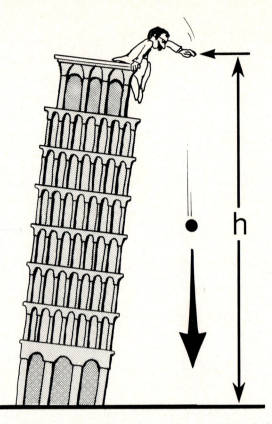

Fig. 2-1 The height of a building can be calculated from Equation 2-1 by timing the falling rock. Can you verify that a 4-second fall means the tower is 78.4 m high?

Answer: Since 6.2 has one digit, 2, to the right of the decimal point, the answer is 69.4.

In multiplication and division, round to the same number of digits as the entry with the least number of significant figures.

Example:
$$4.273$$
$$\times 0.417$$
$$\overline{1.781841}$$

Answer: Since 0.417 has three significant figures (the zero is not significant) and 4.273 has four the answer must have three digits. The answer is 1.78.

Algebra

The rules of algebra provide a method of finding unknown quantities in an equation. Usually these unknowns, also called **variables,** are designated by alphabetic symbols such as x, y, z or certain Greek letters. There are three basic algebraic rules that can be used to solve for the variables in diagnostic ultrasound equations.

Rule 1: When an unknown is multiplied by a number, divide both sides of the equation by that number.

Example: Solve the equation for x: $3x = 10$

Answer: $3x = 10$

$$\frac{3x}{3} = \frac{10}{3} \qquad \textbf{(by rule 1)}$$

$$x = 10/3 = 3\ 1/3$$

Rule 2: When numbers are added to an unknown, subtract that number from both sides of the equation.

Example: Solve the equation $y + 3 = 9$.

$$y + 3 - 3 = 9 - 3 \qquad \textbf{(by rule 2)}$$

$$y = 6$$

Rule 3: When an equation is presented in fractional form, cross multiply and solve for the unknown.

Example: Solve the equation $\dfrac{m}{6} = \dfrac{4}{3}$.

Answer: The crossed arrows show the direction of cross multiplication.

$$\frac{m}{6} \bowtie \frac{4}{3}$$

This yields

$$3m = 24$$

$$\frac{3m}{3} = \frac{24}{3} \qquad \textbf{(by rule 1)}$$

$$m = 8$$

Examples: Solve the following for x.

  a. $4x - 7 = 3$

  b. $\dfrac{3}{x} = \left(\dfrac{2}{3}\right)^2$

  c. $D = C + ABx$

  d. $xy = z$

Answers: a. $4x - 7 = 3$

$$4x - 7 + 7 = 3 + 7 \qquad \textbf{(rule 2)}$$

$$4x = 10$$

$$\frac{4x}{4} = \frac{10}{4} \qquad \textbf{(rule 1)}$$

$$x = 2.5$$

  b. $\dfrac{3}{x} = \left(\dfrac{2}{3}\right)^2$

$$\frac{3}{x} = \frac{4}{9}$$

$$4x = 27 \qquad \textbf{(rule 3)}$$

$$\frac{4x}{4} = \frac{27}{4} \qquad \textbf{(rule 1)}$$

$$x = 6\ 3/4$$

  c. $D = C + ABx$

$$ABx + C = D \qquad \textbf{(rewrite)}$$

$$ABx + C - C = D - C \qquad \textbf{(rule 2)}$$

$$ABx = D - C$$

$$\frac{ABx}{AB} = \frac{(D - C)}{AB} \qquad \textbf{(rule 1)}$$

$$x = (D - C)/AB$$

  d. $xy = z$

$$\frac{xy}{y} = \frac{z}{y} \qquad \textbf{(rule 1)}$$

$$x = z/y$$

Note that examples a and c are nearly identical in form. Symbols such as A, B, C, and D are often used in physics equations instead of numbers.

Scientific notation

Because very large or very small numbers are cumbersome to write, we often use a short form known as power of ten or **scientific notation** to represent such numbers. This system is based on the decimal or power of ten number set. In the quantity 10^4 for example, the 10 represents the base and the 4 is the exponent.

$$10^4 = 10 \cdot 10 \cdot 10 \cdot 10 = 10,000$$

Additional examples of ways to represent numbers in the decimal system are shown in Table 2-1.

To express a number in scientific notation, first write the number in decimal form, accurately locating the decimal point. If there are digits to the left of the decimal point, the exponent will be positive. To determine the value of this positive exponent, position the decimal point after the first digit and count the number of digits the decimal point was moved.

Table 2-1 Various ways to represent numbers in a number system to the base 10

Fractional form	Decimal form	Exponential form
10,000	10,000	10^4
1,000	1,000	10^3
100	100	10^2
10	10	10^1
1	1	10^0
1/10	0.1	10^{-1}
1/100	0.01	10^{-2}
1/1000	0.001	10^{-3}
1/10,000	0.0001	10^{-4}

Example: The distance to the fictional planet Krypton is 21 billion km (21,000,000,000 km). Express this in scientific notation:

Answer: First position the decimal point between the 2 and the 1. Then count the number of digits it was moved. This indicates that the exponent will be $+10$.

$$21,000,000,000 \text{ km} = 2.1 \times 10^{10} \text{ km}$$

If there are no nonzero digits to the left of the decimal point, the exponent will be negative. The value of this negative exponent is found by positioning the decimal point to the right of the first nonzero digit and counting the number of digits the decimal point was moved.

Example: A moustache hair has a diameter of 0.00075 meters (m). What is its diameter in scientific notation?

Answer: First position the decimal point between the 7 and the 5. Next count the number of digits the decimal point has moved, and express this quantity as the negative exponent.

$$0.0075 \text{ m} = 7.5 \times 10^{-4} \text{ m}$$

One common quantity used in physics is the speed of light, symbolized by c. In decimal form:

$$c = 300,000,000 \text{ meters per second (m s}^{-1})$$

Obviously this form is too cumbersome to write each time. Thus the speed of light is always written in scientific notation as:

$$c = 3 \times 10^8 \text{ m s}^{-1}$$

Example: Write the following in scientific notation:
   a. 144000
   b. 425
   c. 0.035
   d. 1/2000

Answer: a. $144000 = 1.44 \times 10^5$
   b. $425 = 4.25 \times 10^2$
   c. $0.035 = 3.5 \times 10^{-2}$
   d. $1/2000 = 0.0005 = 5.0 \times 10^{-4}$

Rules for exponents

The primary advantage of handling numbers in exponential form is evident in operations other than addition and subtraction. The general rules for these types of numerical operations are shown in Table 2-2.

The following examples should emphasize the principles involved.

Example: Simplify the following:
   a. $10^4 \cdot 10^3 / 10^2$
   b. $2^5(2^2/2^8)$
   c. $(5 \times 10^{10})^2$
   d. $(2.718 \times 10^{-4})^3$
   e. $3^2/2^5$

Answer: a. $10^4 \times \dfrac{10^3}{10^2} = 10^{4+3-2} = 10^5$

   b. $2^5 \cdot \dfrac{(2^2)}{(2^8)} = 2^5(2^{2-8}) = 2^5(2^{-6})$
   $$= 2^{5-6} = 2^{-1}$$
   $$= 1/2$$

   c. $(5 \times 10^{10})^2 = 5^2 \times (10^{10})^2$
   $$= 2.5 \times 10^{21}$$

   d. $(2.718 \times 10^{-4})^3 = (2.718)^3 \times (10^{-4})^3$
   $$= 20.08 \times 10^{-12}$$
   $$= 2.008 \times 10^{-11}$$

   e. $\dfrac{3^2}{2^5} = \dfrac{3 \times 3}{2 \times 2 \times 2 \times 2 \times 2} = \dfrac{9}{32}$

Note, as the last example indicates, that the rules for exponents apply only when the bases are the same.

Example: Given $a = 6.62 \times 10^{-27}$, $b = 3.766 \times 10^{12}$, what is
   a. $a \times b$?
   b. $a \div b$?

Answer: a. $a \times b = 6.62 \times 10^{-27} \times 3.766 \times 10^{12}$
   $$= (6.62 \times 3.766) \times (10^{-27} \times 10^{12})$$
   $$= 24.931 \times 10^{-27+12}$$
   $$= 24.93 \times 10^{-15}$$
   $$= 2.49 \times 10^{-14}$$

   b. $a \div b = \dfrac{6.62 \times 10^{-27}}{3.766 \times 10^{12}}$
   $$= (6.62/3.766) \times (10^{-27}/10^{12})$$
   $$= 1.758 \times 10^{-27-12}$$
   $$= 1.76 \times 10^{-39}$$

Binary number system

Computers have become an important part of ultrasound imaging technology. They operate on the sim-

Table 2-2 Rules for handling numbers in exponential form

Operation	Rule	Example
Multiplication	$10^x \times 10^y = 10^{x+y}$	$10^2 \times 10^3 = 10^{2+3} = 10^5$
Division	$10^x \div 10^y = 10^{x-y}$	$10^6 \div 10^4 = 10^{6-4} = 10^2$
Raising to a power	$(10^x)^y = 10^{xy}$	$(10^5)^3 = 10^{5 \times 3} = 10^{15}$
Inverse	$10^{-x} = 1/10^x$	$10^{-3} = 1/10^3 = 1/1000$
Unity	$10^0 = 1$	$3.7 \times 10^0 = 3.7$

Table 2-3 Organization of binary number system

Decimal number	Binary equivalent	Binary number
0	0	0
1	2^0	1
2	$2^1 + 0$	10
3	$2^1 + 2^0$	11
4	$2^2 + 0 + 0$	100
5	$2^2 + 0 + 2^0$	101
6	$2^2 + 2^1 + 0$	110
7	$2^2 + 2^1 + 2^0$	111
8	$2^3 + 0 + 0 + 0$	1000
9	$2^3 + 0 + 0 + 2^0$	1001
10	$2^3 + 0 + 2^1 + 0$	1010
11	$2^3 + 0 + 2^1 + 2^0$	1011
12	$2^3 + 2^2 + 0 + 0$	1100
13	$2^3 + 2^2 + 0 + 2^0$	1101
14	$2^3 + 2^2 + 2^1 + 0$	1110
15	$2^3 + 2^2 + 2^1 + 2^0$	1111
16	$2^4 + 0 + 0 + 0 + 0$	10000

Table 2-4 Power of ten, power of two, and binary notation

Power of ten	Power of two	Binary notation
$10^0 = 1$	$2^0 = 1$	1
$10^1 = 10$	$2^1 = 2$	10
$10^2 = 100$	$2^2 = 4$	100
$10^3 = 1000$	$2^3 = 8$	1000
$10^4 = 10,000$	$2^4 = 16$	10000
$10^5 = 100,000$	$2^5 = 32$	100000
$10^6 = 1,000,000$	$2^6 = 64$	1000000
	$2^7 = 128$	10000000
	$2^8 = 256$	100000000
	$2^9 = 512$	1000000000
	$2^{10} = 1024$	10000000000

plest number system of all, the binary number system. It has only two digits, 0 and 1. The computer performs all operations by converting alphabetic characters, decimal values, and logic functions to binary values. That way, although the binary numbers may become exceedingly long, computation can be handled by properly adjusting the thousands of flip-flop circuits in the computer.

A **bi**nary di**g**it (bit) can have a value of either 0 or 1. Since 2^0 equals 1, the value of the lowest binary number is (0×2^0) or 0. One is represented by (1×2^0) or 1. To represent the decimal numbers 2 and 3, an additional bit is required. The decimal 2 is equal to $(1 \times 2^1) + (0 \times 2^0) = 10$ (read one-zero, not ten). Three is formed by $(1 \times 2^1) + (1 \times 2^0)$ or 11 (one-one) in binary. Four is $(1 \times 2^2) + (0 \times 2^1) + (0 \times 2^0)$ or 100 (one-zero-zero) in binary form. Larger decimal numbers are constructed by

adding appropriate powers of two. For example, the binary number 11111 represents $(1 \times 2^4) + (1 \times 2^3) + (1 \times 2^2) + (1 \times 2^1) + (1 \times 2^0) = 16 + 8 + 4 + 2 + 1 = 31$. As shown in Table 2-3, each time it is necessary to raise 2 to an additional power to express a number, the number of bits required increases by one.

Just as we know the meaning of the power of ten, it is necessary to easily recognize the power of two. Power-of-two notation is used in ultrasound imaging to describe image size, image bit depth (shades of gray), and image storage capacity. Table 2-4 is a review of these power notations.

Example: Convert the binary number 1011010 to decimal form.

Answer: The number contains seven binary digits so 2^0 through 2^6 are required:

$$1011010 = (1 \times 2^6) + (0 \times 2^5) + (1 \times 2^4) + (1 \times 2^3) + (0 \times 2^2) + (1 \times 2^1) + (0 \times 2^0)$$
$$= 64 + 0 + 16 + 8 + 0 + 2 + 0$$
$$= 90$$

Example: Express the number 193 in binary form.

Answer: 193 falls between 2^7 and 2^8. Therefore it will be expressed as 1, followed by seven binary digits.

$$193 = (1 \times 2^7) + (? \times 2^6) + (? \times 2^5) + (? \times 2^4) + (? \times 2^3) + (? \times 2^2) + (? \times 2^1) + (? \times 2^0)$$

By trial and error one can determine the missing numbers:

$$193 = (1 \times 2^7) + (1 \times 2^6) + (0 \times 2^5) + (0 \times 2^4) + (0 \times 2^3) + (0 \times 2^2) + (0 \times 2^1) + (1 \times 2^0)$$
$$193 = 128 + 64 + 1$$
$$\text{or } 193 = 11000001$$

In computer language, a single binary digit, 0 or 1, is called a bit. The computer will employ as many bits as necessary to express a decimal digit. The 26 characters of the alphabet and other special characters are usually encoded by 8 bits. To **encode** is to translate from ordinary characters to computer-compatible characters or binary digits. Depending on the microprocessor, a string of 8, 16, or 32 bits will be manipulated simultaneously.

Bits are often grouped into bunches of eight called **bytes.** Computer capacity is expressed by the number of bytes that can be accommodated by computer memory. The most popular personal computers employ 8- and 16-bit microprocessors with 64 to 512 kilobytes of memory. One kbyte is equal to 8×1024 bits. Note that kilo is not metric in computer use. Instead it represents 2^{10}. The minicomputers used in radiology have capacities measured in megabytes, where 1 Mbyte = 1 kbyte $\times$ 1 kbyte = $2^{10} \times 2^{10} = 2^{20} = 1,048,576$ bytes.

Example: How many bits can be stored on a 64-byte chip whose byte size is 8 bits?

Answer: 1024 bits/kbytes $\times$ 64 kbytes $\times$ 8 bits = $2^{10} \times 2^6 \times 2^3 = 2^{19} = 524,288$

Depending on the computer configuration, 2 bytes usually constitute a word. In the case of a 16-bit microprocessor, a word would be 16 consecutive bits of information that are interpreted and shuffled about the computer as a unit. Each word of data in memory has its own address.

Graphing

Most graphs consist basically of two axes: a horizontal, or x axis, and a vertical, or y axis. The point where the two axes meet is called the origin (labeled 0 in Fig. 2-2). Coordinates have the form of ordered pairs (x,y), where the first number of the pair represents a distance along the x axis, and the second number indicates displacement in the y direction. For example, the ordered pair (3,2) represents a point three units over on the x axis and up two units on the y axis. This point is plotted in Fig. 2-2. How does it differ from the point (2,3)? If the value of one additional ordered pair is known, a straight line can be drawn between the points.

In physics the axes of graphs are not usually labeled x and y. Generally, the relationship between two specific quantities is desired.

Example: Graph the relative pressure of an ultrasound beam emitted from a transducer versus distance from the transducer given the following data:

Distance from transducer (cm)	Relative pressure
1	.25
2	.50
3	.70
4	.85
6	1.00
8	.90
10	.67
12	.45

Answer: The first step is to draw the axes on graph paper. In this example the data are recorded in ordered pairs, where distance represents the x value and relative pressure represents the y value. Next, note the range of each quantity, and choose a convenient scale that allows the graph to adequately fill the page. Very small graphs should generally be avoided. Then label the axes and carefully plot each point. Finally, draw the best smooth curve through the points. The curve need not touch each of the plotted points. The completed graph is shown in Fig. 2-3.

UNITS OF MEASUREMENT

Physicists strive for certainty and simplicity. Thus only three measureable quantities are set aside as the basis of all others. These **base quantities** are length, mass, and time. Fig. 2-4 indicates the role these base quantities play in supporting some of the other quantities used in physics. The secondary quantities are called **derived quantities,** derived from a combination of one or more of the three base quantities. For example, volume is length cubed (1^3), density

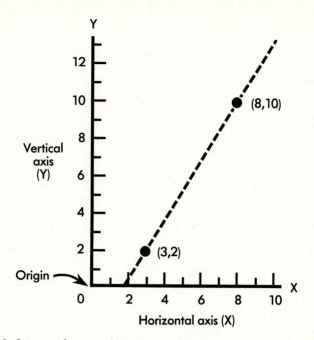

Fig. 2-2 Principle features of any graph are the x and y axes that intersect at the origin. Points of data are entered in ordered pairs.

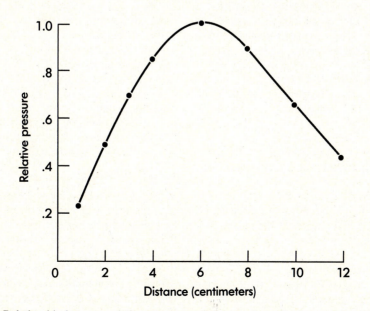

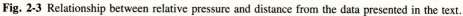

Fig. 2-3 Relationship between relative pressure and distance from the data presented in the text.

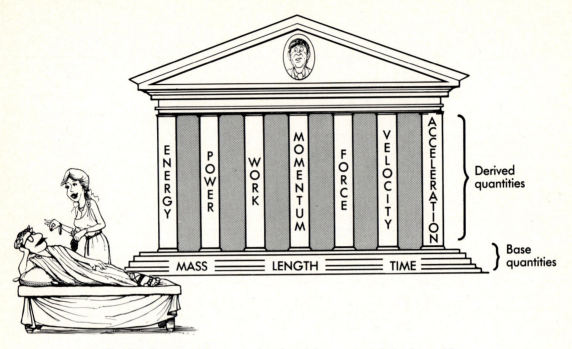

Fig. 2-4 Base quantities, mass, length and time, support derived quantities.

is mass divided by volume ($m\ l^{-3}$), and velocity is length divided by time ($l\ t^{-1}$). There are additional special quantities, built on the derived quantities, designed to support measurement in specialized areas of science and technology.

Whether a physicist is studying something large, such as the universe, or a small object, such as an atom, meaningful measurements must be reproducible. Therefore once the fundamental quantities are established, it is essential that they be related to a well-defined and invariable standard. Standards are normally defined by international organizations and usually redefined when the progress of science requires greater precision.

Standard of length

For many years the standard unit of length was accepted to be the distance between two lines engraved on a platinum-iridium bar kept at the International Bureau of Weights and Measures in Paris, France. This distance was defined to be exactly one meter (m). A meter is about three inches longer than a yard and is subdivided into the centimeter (100 cm =

1 m) and the millimeter (10 mm = 1 cm).

The exact relationships between the British and metric system of units are:

1 yd = 0.9144 m

1 in = 2.54 cm = 25.4 mm

In 1960 the need for a more accurate standard of length led to the redefinition of the meter in terms of the wavelength of orange light emitted from an isotope of krypton (^{86}Kr). One meter is now defined as 1,650,763.73 wavelengths of this light.

Standard of mass

In the same vault in Paris where the standard meter is kept, there is a cylinder of platinum-iridium that represents the standard of mass—the kilogram (kg). The kilogram was defined to represent the mass of 1000 cm^3 of water at 0° Celsius (C). The kg is subdivided into grams (g) (1000 gm = 1 kg). Mass is not the same quantity as weight; the kilogram is a unit of mass, whereas the newton or the pound, a British unit, is a unit of weight.

Table 2-5 Systems of units

	MKS*	CGS	British
Length	Meter (m)	Centimeter (cm)	Foot (ft)
Mass	Kilogram (kg)	Gram (g)	Pound (lb)†
Time	Second (s)	Second (s)	Second (s)

*The more general SI system includes four additional base units.
†The pound is actually a unit of force but is related to mass.

Standard of time

The standard unit of time is the second (s). Originally the second was defined in terms of the rotation of the earth on its axis—the mean solar day. In 1956 it was redefined to be a certain fraction of the tropical year 1900. In 1964 the need for a better standard of time led to another redefinition. Now time is measured by an atomic clock and is based on the vibrations of cesium atoms. The atomic clock is capable of keeping time correctly to about 1 second in 5000 years. The accuracy of future hydrogen maser clocks promises to be even greater, perhaps to 1 second in several million years!

Systems of units

Every measurement has two parts: a magnitude and a unit. For example, if a book has a length of 25 cm, it makes no sense to report only the magnitude, 25, without designating a unit. Here the unit of measurement is the centimeter.

Table 2-5 shows that there are three systems of units to represent the base quantities. The MKS (meters, kilograms, seconds) and the CGS (centimeters, grams, and seconds) systems are more widely used in science and in most countries of the world than is the British system. Le Systeme International d'Unites (The International System, SI) is an extension of the MKS system and represents the present state of the art for units. SI includes the three base units of the MKS system plus an additional four. There are derived units and special units of SI to represent derived quantities and special quantities.

Rule: The same system of units must always be used when working problems or reporting answers.

Example: What is unacceptable about the following?

a. Density = 8.1 lb m^{-3}

b. Mass per unit area = 700 g m^{-2}

Answer: Density should be expressed as kg m^{-3} or in g cm^{-3}. Mass per unit area should be expressed as kg m^{-2} or g cm^{-2}.

Example: The dimensions of a rectangular box are found to be 30 cm × 86 cm × 4.2 m. Find the volume.

Answer: The formula for the volume of a rectangle is given by

V = length × width × height

or

V = lwh

However, since the dimensions are given in different systems of units, we must choose only one system. Therefore

V = (0.30 m)(0.86 m)(4.2 m)
  = 1.1 m³

Note that the units are multiplied also:

m × m × m = m³

Example: Find the density of a ball with a volume of 200 cm³ and a mass of 0.4 kg.

Answer: Change 0.4 kg to 400 g

D = mass/volume
  = 400 g/200 cm³
  = 2 g cm^{-3}

SCIENTIFIC PREFIXES

Often in ultrasound, very large or very small multiples of base or derived units must be described. For example, a one-million Hertz (Hz) ultrasound transducer can be written as 1 MHz by using the prefix M (Mega) to represent one million. Table 2-6 presents a list of prefixes and symbols commonly used in scientific application. As shown in the examples below, it is often necessary to change from the standard scientific notation format to get the correct power of ten so that the appropriate prefix can be substituted.

Table 2-6 Standard scientific and engineering prefixes

Multiple	Prefix	Symbol
10^9	giga-	G
10^6	mega-	M*
10^3	kilo-	k*
10^2	hecto-	h
10	deka-	da
10^{-1}	deci-	d*
10^{-2}	centi-	c*
10^{-3}	milli-	m*
10^{-6}	micro-	μ*
10^{-9}	nano-	n
10^{-12}	pico-	p

*Indicates most commonly used prefixes.

Example: The velocity of ultrasound in a certain material is

19,000 m s^{-1}. Express this as km s^{-1}

Answer: 19,000 m s^{-1} = 19 × 10^3 m s^{-1}
= 19 k ms^{-1}

Example: Convert the following:
a. 0.0015 m to mm
b. 2,700,000 Hz to MHz
c. 0.0018 s to μs
d. .005 m to cm

Answer: a. 0.0015 m = 1.5 × 10^{-3} m = 1.5 mm
b. 2,700,000 Hz = 2.7 × 10^6 Hz = 2.7 MHz
c. 0.00018 s = 1.8 × 10^{-4} s = 180 × 10^{-6} s = 180 μs
d. 0.005 m = 5.0 × 10^{-3} m = 0.5 × 10^{-2} m = 0.5 cm

Review Questions: Chapter 2

1. What is the decimal equivalent of the following?
   a. 7/10
   b. 81/1000
   c. 4/100000
   d. 7/15
   e. 3/8

2. Write in scientific notation:
   a. 5,630,000,000
   b. 0.00000092
   c. 941
   d. 0.253
   e. $(2 \times 10^4)/(8 \times 10^7)$
   f. $(5 \times 10^{15})/(1.5 \times 10^{12})$

3. Solve the following for x:
   a. $3x - 6 = 9$
   b. $3/x = 4/5$
   c. $7(3 - x) = 8$
   d. $(x/4)^2 = (3/2)^3$
   e. $3BQx = 9TV$

4. Remove the prefix and express in standard scientific notation.
   Example: 12.5 MHz = 12.5 × 10^6 Hz = 1.25 × 10^7 Hz
   a. 140 km =   meters =   meters
   b. 75 μV =   volts =   volts
   c. 300 ms =   seconds =   seconds
   d. 6.05 cg =   grams =   grams

5. Express each with a scientific prefix:
   Example: 400 × 10^{-7}A = 40 × 10^{-6}A = 40μA
   a. 7,000,000 Hz =
   b. 1.6 × 10^{-3} s =
   c. 0.00018 g =
   d. 44,000 m =
   e. 0.004 m =

6. Express the following in decimal form:
   a. 110
   b. 1011
   c. 11101

7. Express the following in binary form:
   a. 7
   b. 18
   c. 237

8. Identify the three base quantities and the SI unit for each quantity.

9. Determine the mass and length of your shoe in the MKS system, the CGS system, and the British system.

3 Basic Physical Quantities

Mechanics is a segment of physics that deals with motion. Diagnostic ultrasound uses high frequency mechanical waves transmitted through and reflected from body tissues. Doppler ultrasound uses the motion of blood in the heart or vessels to construct dazzling multicolor images of flow. Thus, the study of mechanics is important to understanding the physics of ultrasound.

The motion of an object can be described by two terms: velocity and acceleration. Velocity, sometimes called speed, is a measure of how fast something is going. The speedometer on your car measures the velocity of a car in miles per hour (or kilometers per hour). In ultrasound, velocities are given in meters per second.

VELOCITY

Definition:	Velocity is the rate of change of the position of an object with time.
Symbol:	v
Units:	meters per second (m s^{-1})
Equation:	$v = d/t$ **(3-1)**
	In this equation v represents average velocity and d is distance traveled in time t.

Example: What is the velocity of blood in a vessel that travels 6 cm in 4 s?

Answer: $v = d/t$
$= 6 \text{ cm}/4 \text{ s}$
$= 1.5 \text{ cm s}^{-1} \text{ or } 0.015 \text{ m s}^{-1}$

Example: If the state champion swimmer, shown in Fig. 3-1, can swim at the rate of 2.04 m s^{-1}, how long will it take him to swim 50 m?

Answer: $v = d/t$
$t = d/v$
$t = 50\text{m}/2.04 \text{ m s}^{-1}$
$t = 24.5 \text{ s}$

ACCELERATION

Definition:	Acceleration is the rate of change in velocity with time; that is, how "fast" the velocity is changing.
Symbol:	a
Units:	meters per second squared (m s^{-2})
Equation:	$a = v/t$ **(3-2)**

If velocity is constant, the acceleration is zero. A constant acceleration of 2 m s^{-2} means that the velocity of an object increases by 2 m s^{-1} each second.

NEWTON'S LAWS OF MOTION

In the year 1686 English mathematician Sir Issac Newton presented three principles that, even today, are recognized as fundamental laws of motion.

First law

Newton's first law states that a body will remain at rest or continue moving with a constant velocity in a straight line unless acted on by an external force. This law says that if no force acts on an object, there will be no acceleration. The property of matter that acts to resist a change in its state of motion is called **inertia.** Newton's first law is thus often referred to as the Law of Inertia. Fig. 3-2 illustrates this principle.

A portable ultrasound machine will not move until forced by a push or pull. However, once in motion it would continue to move forever, even when the pushing force is removed, except that an opposing force, friction, is present.

Second law

Newton's second law, illustrated in Fig. 3-3, is a definition of the concept of force.

Definition:	Force can be thought of as a push or pull on an object.
Symbol:	F

Fig. 3-1 Maximum velocity of a world-class swimmer is about 2 m s^{-1}; velocity of ultrasound is 1540 m s^{-1} in soft tissue.

Units: SI: Newton (N) (1N = 1 kg m s^{-2})
CGS: dyne (1 dyne = 1 g cm s^{-2})

Equation: F = ma   (3-3)

If a body of mass **m** has an acceleration **a,** then the force on it is given by the mass times the acceleration.

Example: Find the force on a 34 kg mass accelerated at 4 m s^{-2}

Answer: F = ma
= (34 kg)(4 m s^{-2})
= 136 N

Third law

Newton's third law of motion states that to every action there is an equal and opposite reaction. According to this law, if a heavy block is pushed, the block will push back with the same force that is applied. If a very large force is exerted, the block will begin to accelerate away, continuing to equalize the interaction (Fig. 3-4).

WORK

Definition: Work, as used in physics, has a specific meaning. The work done on an object is the force applied times the distance over which it is applied.

Symbol: Work

Units: SI: Joules (J); CGS: ergs
Conversion: 1 J = 10^7 ergs

Equation: Work = F · d   (3-4)

When a sonographer lifts a transducer he or she is doing work. However, when the transducer is merely held motionless, no work is being performed, even though some effort is expended in holding it there.

Example: Find the work done in lifting a transducer weighing 2 N to a height of 0.3 m.

Answer: Work = F · d
= (2 N)(0.3 m)
= 0.6 J

Fig. 3-2 Practical illustration of Newton's first law.

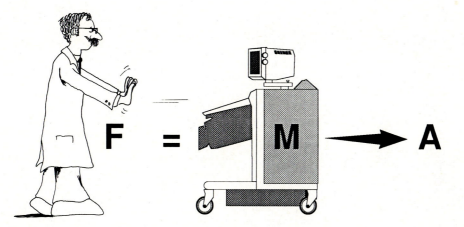

Fig. 3-3 Newton's second law states that the force applied to move an object is equal to the mass of the object times the object's acceleration.

POWER

Definition: Power is the rate of doing work.

Symbol: P

Unit: Watt

Equation: P = Work/t   (3-5)

Recall that the same amount of work is required to lift a transducer to a given height, whether it takes 1 second or 1 year to do so. Power gives us a way to include the time required to perform the work.

Example: A sonographer lifts the 2N transducer from the preceding example to 0.3 m in 0.4 s. What is the power exerted?

Answer: P = Work/t
= 0.6 J/0.4 s
= 1.5 W

ENERGY

Definition: Energy is the ability to do work.

Symbol: E

Fig. 3-4 Illustration of Newton's third law after midterm grades have been issued. Physics instructor pushes with all available force on walls being closed because of greater force applied by disgruntled students.

Units: SI: Joules (J);CGS: ergs

The law of conservation of energy states that enaergy may be transformed from one form to another, but it cannot be created or destroyed; the total energy is constant. For example, electrical energy is converted to ultrasound energy in an ultrasound transducer.

There are many forms of energy including mechanical, chemical, electrical, thermal, and nuclear.

Mechanical energy

There are two forms of mechanical energy—**kinetic** and **potential.** Kinetic energy (KE) is associated with the motion of an object. From the defining equation

$$KE = 1/2\ mv^2 \qquad \text{(3-6)}$$

where m = mass (kg)
v = velocity (m s^{-1})

it is apparent that kinetic energy depends on the mass of the object and on the square of its velocity.

Potential energy

Potential energy (PE) is the stored energy of position or configuration. A book on a table has PE because of its height above the earth. It has the ability to do work by falling back to the ground. Gravitational potential energy can be calculated from the equation:

$$PE = mgh \qquad \text{(3-7)}$$

where h is the distance above the earth's surface g is the acceleration of gravity on earth.

In the MKS system g = 9.8 m s^{-2}.

A coiled spring and a stretched rubber band are examples of other systems that have potential energy because of their unstable configurations.

Chemical energy

Chemical energy is energy released by way of a chemical reaction. An example is the energy provided to our bodies through chemical reactions involving the food we eat. At the molecular level this area of science is called **biochemistry.** The energy released when a stick of dynamite explodes is a more dramatic example of chemical energy.

Electrical energy

Electrical energy represents the work done when an electron or an electronic charge moves through an electric potential. The most familiar form of electrical energy is normal household electricity, which involves the movement of electrons through a copper wire under an electric potential of 110 volts (V). All electric apparatus, such as motors, TVs, and ultrasound transducers, function through the use of electrical energy.

Thermal energy

Thermal (heat) energy is energy of motion at the atomic or molecular level and may be viewed as the

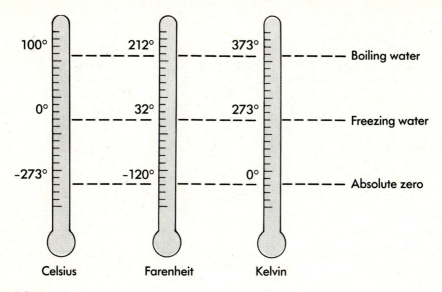

Fig. 3-5 Three principal scales used to represent temperature. Kelvin is the universally adopted scale for scientific purposes.

kinetic energy of atoms. Thermal energy is measured by temperature. The faster the atoms and molecules of a substance are moving, the more heat energy the substance contains, and the higher its temperature. Therapeutic ultrasound uses intense ultrasonic energy to raise the temperature of tissues in the body.

Nuclear energy

Nuclear energy is energy contained in the nucleus of an atom. The release and use of this type of energy is controlled in nuclear electric-power-generating plants. An example of the uncontrolled release of nuclear energy is the atomic bomb.

HEAT

The standard definition of heat is the random disordered motion of molecules. The more rapid and disordered the motion, the more heat the body contains. Energy absorbed from an ultrasound beam is converted to heat, as discussed later. The unit of heat, the calorie, is defined as the heat necessary to raise the temperature of 1 g of water 1° C. The same amount of heat will have different effects on different materials. For example, the heat required to change the temperature of 1 g of silver by 1° C is approximately 0.05 calorie, or only one twentieth that required for a similar temperature change in water.

Heat is transferred from one place to another in three ways:

1. **Conduction** is the transfer of heat by molecular motion from a high temperature. Conduction is easily observed when a hot and cold object are placed on contact. After a short time, heat conducted to the cooler object will result in equalization of temperatures.
2. **Convection** is the mechanical transfer of "hot" molecules in a gas or liquid from one place to another. A steam radiator or a forced-air furnace warms a room by convection. The air around the radiator is heated, causing it to rise, while cooler air circulates in to replace it. A forced-air furnace blows heated air into the room, providing forced circulation to complement the natural convection.
3. **Radiation** is a method of heat transfer that depends on the temperature of the object. The reddish glow emitted by hot objects is evidence of heat transfer by radiation.

Temperature is normally measured with a reproducible scale called a thermometer. A thermometer is usually calibrated at two reference points: the freezing and boiling points of water. Fig. 3-5 shows the relationship of three scales that have been developed to measure temperature: Fahrenheit (F), Celsius (C), and Kelvin (K).

These scales are interrelated by the following equations, where the subscripts c, f, and k refer to Celsius, Fahrenheit, and Kelvin, respectively.

$$T_c = 5/9 \ (T_f - 32) \tag{3-8}$$

$$T_f = 9/5 \ T_c + 32 \tag{3-9}$$

$$T_k = T_c + 273 \tag{3-10}$$

Example: Convert 98° F to degrees Celsius.

Answer: $T_c = 5/9 \ (T_f - 32)$
$\qquad = 5/9 \ (98 - 32) = 5/9 \ (66) = 36.7° \ C$

MECHANICAL EQUIVALENT OF HEAT

In about 1840 James Joule performed an ingenious experiment to show that heat is a form of energy. He used a falling weight to turn a set of paddles in a water container. The energy given up by the weight was shown to be equivalent to the heat gained by the water. This helped to demonstrate the conservation of energy principle, and the SI unit of energy was named in his honor. The currently accepted value for the mechanical equivalent of heat is 1 calorie = 4.186 Joules.

Review Questions: Chapter 3

1. Define or otherwise identify the following and give their MKS units.
    a. Velocity
    b. Acceleration
    c. Force
    d. Work
    e. Power
    f. Energy
    g. Heat

2. Ultrasound travels through a 3 m thick section of a certain material in 2 milliseconds (ms). Calculate the velocity of ultrasound in this material.

3. How long does it take ultrasound to travel through a 1 mm thick crystal if the velocity of ultrasound in the crystal is 4000 m s^{-1}.

4. If an ultrasound wave propagates through tissue with a velocity of 1540 m s^{-1}, how long will it take the wave to traverse .01 m? Express your answer in μs.

5. An ultrasound scanner is calibrated so that the average velocity of propogation is assumed to be that in soft tissue, 1540 m s^{-1}. The velocity in bone however is more than twice this value. Assume the ultrasound beam traverses 1 cm of bone followed by 1 cm of soft tissue which contains a small tumor. Discuss whether the tumor will appear to be closer or further from the transducer than its true 2 cm depth.

6. How much work is done in lifting an object weighing 2.6 N to a height of 2.6 m? What power is exerted if the time required is 0.3 s?

7. Water has a mass density of 1 g cm^{-3}. Convert this density to MKS units.

8. One dyne is equal to 1 g cm s^{-2}. How many dynes are in a Newton?

9. Compute the number of ergs in a joule if 1 erg equals 1 g cm^2 s^{-2}.

10. Briefly discuss the three methods of heat transfer.

11. Match the following:
1. Energy	a. ability to do work
2. Power	b. a push or a pull
3. Work	c. rate of doing work
4. Force	d. force times distance
5. Kinetic energy	e. energy of motion
6. Potential energy	f. energy of position

12. How could you show that heat is a form of energy?

4 Fundamentals of Waves

Like the surfer in Fig. 4-1, sonographers encounter many types of waves in everyday life. Radiowaves travel at the speed of light and do not need a medium for propagation. Ocean waves travel much slower and are produced by the moon's gravitational tug on the earth. The sound waves of human speech are produced by mechanical deformation of air caused by vibrations of the vocal chords in the larynx. The first is an example of electromagnetic waves while the last two are examples of mechanical waves.

ELECTROMAGNETIC WAVES

Ever present around us is a field or state of energy called electromagnetic radiation. The electromagnetic spectrum is composed of many different types of radiation including gamma rays, x-rays, ultraviolet light, visible light, infrared light, microwaves, and radiowaves. Each of these forms of electromagnetic radiation travels at the speed of light (c = 3×10^8 m s^{-1}) and does not need a medium for propagation. Electromagnetic waves are distinguished by their energy, frequency, and wavelength. X-rays, for example, have a shorter wavelength and higher energy than microwaves. However, all types of electromagnetic radiation are fundamentally the same. Each can be represented as a bundle of energy consisting of electric and magnetic waves traveling at the speed of light.

MECHANICAL WAVES

Mechanical waves, defined as the propagation of energy through a medium by cyclic pressure variation, need a deformable, elastic medium, like air, water, or soft tissue for propagation. It is easy to show this for **sound** since our ears serve as built-in detectors. If a ringing alarm clock is placed in a bell jar and the air is evacuated, the ringing sound grows fainter and fainter as the air is removed. Finally, as shown in Fig. 4-2, when there is not enough air in the jar to transmit the sound, the alarm will no longer be heard. However, **light** is readily transmitted through the evacuated glass jar since electromagnetic radiation does not need a medium for propagation. Table 4-1 contrasts the characteristics of electromagnetic radiation and sound/ultrasound.

Pressure variations in the mechanical wave cause displacement of particles in the medium and oscillation of these particles about their equilibrium positions. The wave travels through the medium by the progressive interaction of individual particles as they oscillate back and forth. The energy in the mechanical wave is transmitted through the motion of these particles. The medium itself does not move from one place to another.

Ultrasound behaves as a mechanical wave. The ultrasound beam may be considered either a displacement of particles in the medium or the pressure incident to the displacement of these particles. When the displacement from the equilibrium point of a particle is at a maximum or minimum the pressure there is zero. When the displacement of the particle is zero, the pressure there is at a maximum. The displacement caused by the highest peak pressure from a phased array transducer in soft tissue is about 10^{-8} m. This displacement represents about 100 times the diameter of an atom, which is approximately 10^{-10} m.

Mechanical waves are categorized by the direction of the displacement of individual particles in the medium, either transverse or longitudinal.

Transverse waves

In a transverse wave, particles vibrate perpendicular to the direction of the wave motion. Fig. 4-3 illustrates the anatomy of a transverse wave in water. If a cork was placed in a pond before a rock was thrown in, the cork would simply vibrate up and down in

Fig. 4-1 World famous surfer prepares to use ocean waves (mechanical waves) for ultimate surfing trick called "hang it up" while radio broadcast of event (electromagnetic waves) thrills the world.

Fig. 4-2 Sound needs a medium for propagation. If the medium *air* is removed from the bell jar, the alarm is no longer heard. However *light,* which consists of electromagnetic waves, is readily transmitted through the vacuum as shown by the distorted face peering through the jar.

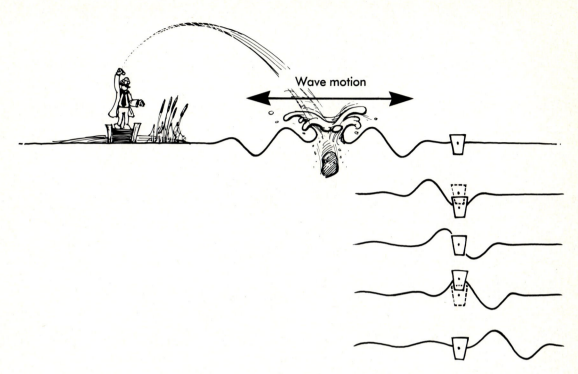

Fig. 4-3 Example of a transverse wave. When a rock is thrown into a still pond, a cork in the water will move up and down or perpendicular to the direction of the wave motion.

Table 4-1 Comparison of electromagnetic radiation and sound/ultrasound

Characteristic	Electromagnetic radiation	Sound/ultrasound
Velocity	Speed of light $c = 3.0 \times 10^8$ m s^{-1}	Depends on medium Velocity in air = 330 m s^{-1} Velocity in tissue = 1540 m s^{-1}
Propagation	Transverse wave—no medium required, vacuum or medium	Longitudinal wave—medium required, solid, liquid, or gas
Frequency	Wide variation: from radio—100 Hz to gamma rays—10^{24} Hz	Sound 20-20,000 Hz US > 20,000 Hz Dx US—1-10 MHz
Production	Vibrating electrical charge	Vibrating source: sound—larynx US—piezoelectric crystal

the same place every time a wave passed. This is because the individual droplets of water in the pond also vibrate up and down, transverse to the direction of the wave motion.

Although not mechanical waves, electromagnetic waves such as light, x-rays, and radiowaves are also examples of transverse waves. As shown in Table 4-1, electromagnetic waves are not mechanical waves since they do not need a medium for propagation and travel many orders of magnitude faster

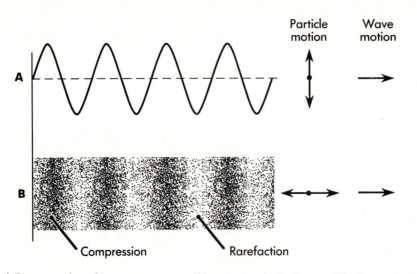

Fig. 4-4 Representation of a transverse wave **(A)** and a longitudinal wave **(B)**. The peak of the transverse wave corresponds to a region of compression in the longitudinal wave.

than mechanical waves. Fig. 4-4, *A* illustrates a representation of the transverse wave. The wavelength is shown as the peak to peak distance.

Longitudinal waves

Acoustic waves, which include infrasound, sound, and ultrasound, are examples of longitudinal waves. As Fig. 4-4, *B* shows, the acoustic wave consists of areas of high and low pressure. A high pressure region (**compression**) is called a peak or anti-node, whereas areas of low pressure (**rarefaction**) are labeled valleys or nodes. One wavelength is the distance between two adjacent areas of compression or rarefaction. Particles in the medium move parallel to the direction of wave travel.

Even though it is technically incorrect, ultrasound waves are often represented for simplicity like the transverse wave shown in Fig. 4-4, *A*. Although use is made of this model to simplify the presentation of the ultrasound parameters, remember that the actual ultrasound beam consists of a longitudinal wave similar to that shown in Fig. 4-4, *B*.

To visualize how an ultrasound beam is propagated, imagine that the molecules are connected by springs as shown in Fig. 4-5. An ultrasound wave incident on the first molecule provides a force that will compress the spring between the first and second molecules. This compression is transferred in turn between molecules two and three and in between

pairs of neighboring molecules until friction eventually will cause the molecular motion to dissipate and the wave ceases to exist. Each of the molecules oscillates about its equilibrium position every time a wave passes. In acoustics, this variation of pressure with time that causes the atoms to oscillate is the variable most often used to describe the wave.

Surface waves

Some waves are not simply longitudinal or transverse. Such waves are classified as **surface waves.** With surface waves, the particle movement is restricted to a thin layer at the surface of the medium supporting the waves.

WAVE PROPERTIES

There are several descriptive features of waves that require further understanding.

Wavelength

Definition:	The wavelength of an acoustic wave is the distance between two adjacent bands of compression or rarefaction.
Symbol:	λ (Greek letter Lambda)
Units:	meters (m)

Using the transverse wave model shown in Fig. 4-6, wavelength is represented as the peak to peak or trough to trough distance, or the distance between

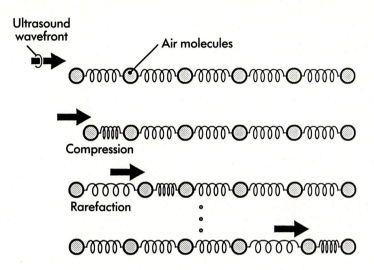

Fig. 4-5 Idealistic representation of sound propagating through air molecules connected with springs. Sound is a longitudinal wave—the air molecules vibrate in the direction of wave motion.

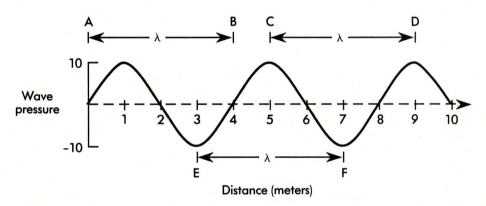

Fig. 4-6 Wavelength (λ) is the distance within a wave that includes a positive and negative cycle (**A** to **B**). Wavelength can also be represented as the distance between adjacent peaks (**C** to **D**) or valleys (**E** to **F**). In each case the wavelength shown here is 4 m.

a positive and negative cycle. Note that wavelength is a **length** and the SI unit is meters. The wavelength produced by a 1.5 MHz transducer in soft tissue is about 1 mm.

Frequency

Definition:	Frequency is the number of wavelengths (cycles) that pass a given point in 1 second.
Symbol:	f

Units:	Hertz (Hz); Note: 1 Hz = 1 cycle/s = $1\ s^{-1}$ (cycles are dimensionless)

The unit of frequency is named for the physicist Heinrick Hertz, an early researcher in the field of sound. As discussed in Chapter 1, sound includes the range of frequencies that are audible to the human ear—20 to 20,000 Hz. Ultrasound includes longitudinal waves above 20,000 Hz. Diagnostic ultrasound transducers emit ultrasound radiation with frequencies ranging from 1 to 10 MHz. The ultrasound frequency used in imaging is important for resolution and the depth of penetration in the patient. Chapter 9 will discuss how axial resolution improves with increasing frequency, but the depth of beam penetration decreases as frequency increases.

AM radio stations produce electromagnetic radiowaves with frequencies in the kilohertz (kHz) range. As mentioned earlier, radiowaves are fundamentally different from ultrasound but the same units of frequency (Hz) are used for both.

Period

Definition:	Period is the time required for one complete cycle.
Symbol:	T
Units:	Seconds (s)

Period and frequency have an inverse relationship given by:

$$T = 1/f \qquad (4\text{-}1)$$

Example: If the frequency is 10 MHz, what is the period?

Answer: $T = 1/f$
$T = 1/(10 \times 10^6 \text{ s}^{-1})$
$T = 10^{-7} \text{ s} = 10 \text{ μs}$

Velocity

Definition:	Velocity is the speed at which the wave moves through the medium. It is often called the speed of propagation.
Symbol:	v
Units:	meters per second (m s^{-1})

The velocity of all electromagnetic radiation equals the speed of light (c = 3.0×10^8 m s^{-1}). For ultrasound and other acoustic waves, velocity depends on the medium through which the beam is transmitted. Table 4-2 lists the velocity of ultrasound in various materials. Note that air transmits ultrasound with the lowest velocity and bone with the highest velocity in biological materials. The velocity of ultrasound in soft tissue is 1540 m s^{-1}.

All acoustic waves are transmitted through the same medium at the same velocity even though their frequencies are different. That is, a 1 MHz ultrasound beam and a 10 MHz beam travel through soft tissue at the same velocity, 1540 m s^{-1}. A sim-

Table 4-2 Velocity of ultrasound in various materials (m s^{-1})

Biological material		Nonbiological material	
Air	330	Mercury	1450
Fat	1450	Castor oil	1500
Water	1540	PZT	4000
Soft tissue	1540	Steel	5850
Blood	1570		
Muscle	1585		
Skull (bone)	4080		

ilar relationship holds for sound waves. Imagine the confusion that would arise at a symphony if the waves from the high frequency piccolo arrived at your ears before those from the low frequency tuba.

Wave equation

One of the most important equations in diagnostic ultrasound relates the frequency, wavelength, and velocity. The equation is called The Wave Equation and is given by:

$$v = f\lambda \qquad (4\text{-}2)$$

where v = the velocity of sound in the medium (m s^{-1})
f = the frequency (Hz), and
λ = the wavelength in m.

Example: Assume a transducer operates at 2 MHz. What is the wavelength in soft tissue?

Answer: $v = f\lambda$, therefore
$\lambda = v/f$
$\lambda = 1540 \text{ m s}^{-1} \div 2 \times 10^6 \text{ s}^{-1}$
$\lambda = 0.77 \times 10^{-3} \text{ m} = 0.77 \text{ mm}$

Example: A diagnostic ultrasound beam has a wavelength of 2 mm in the transducer element where the velocity is 4000 m s^{-1}. What is the generating frequency?

Answer: $v = f\lambda$, therefore
$f = v/\lambda$
$f = (4000 \text{ m s}^{-1}) \div 2 \times 10^{-3} \text{m}$
$f = 2 \times 10^6 \text{ s}^{-1}$
$f = 2 \text{ MHz}$

Several points should be clear from the previous examples and discussion:

1. The two media in the examples, transducer element (crystal) and soft tissue, transmit ultrasound with different velocities: $\lambda_{crystal}$ = 4000 m s^{-1} and λ_{tissue} = 1540 m s^{-1}.

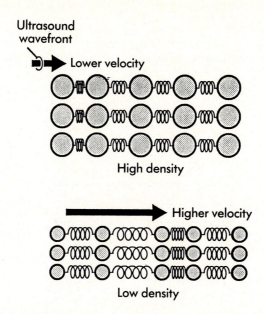

Fig. 4-7 Ultrasound beam incident on simulated molecular structures of high and low density. The ultrasound velocity is lower after interaction with high density molecules.

Table 4-3 Relationship of velocity, frequency, and wavelength in soft tissue and transducer element

Material	Velocity (m s^{-1})	Frequency (MHz)	Wavelength (mm)
Soft tissue	1540	0.77	2.0
Soft tissue	1540	2.0	0.77
Crystal	4000	5.2	0.77
Crystal	4000	0.77	5.2

2. The wavelengths of ultrasound in the two media are different; $\lambda_{crystal}$ = 2 mm and λ_{tissue} = 0.77 mm for a 2 MHz beam. Thus, when the same frequency is employed in different materials, the wavelength is longer in the material with the higher velocity.

3. If only one medium, such as tissue, is considered, the velocity of ultrasound in the medium is a constant 1540 m s^{-1} regardless of the frequency employed. Thus in a single medium, according to the wave equation, the product of frequency and wavelength is a constant. This means that as frequency increases, wavelength must decrease. Table 4-3 illustrates this relationship. A 2.0 mm wavelength can be produced in soft tissue by a 0.77 MHz frequency. Likewise, a 0.77 mm wavelength can be produced in tissue by a 2.0 MHz frequency. Table 4-3 shows that the same relationship holds for

a transducer crystal although it requires a 5.2 MHz frequency to produce a 0.77 wavelength.

Density and compressibility: effects on velocity

Density. Rule: **The more dense the material, the slower the velocity of ultrasound.** To understand this, think of large and small molecules connected with springs as shown in Fig. 4-7. An ultrasound beam incident on each of these would find the large, more dense molecules much harder to move because the larger molecules have greater inertia. Therefore, the velocity of ultrasound would tend to be lower after interacting with large dense molecules than when interacting with small more easily moved molecules.

Compressibility. Rule: **The more compressible the material, the slower the velocity of ultrasound.** Compressibility is the fractional change in a volume of material caused by a change in pressure.

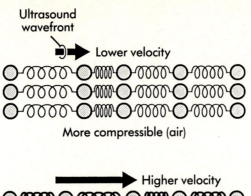

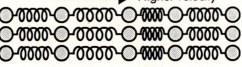

Fig. 4-8 Ultrasound velocities are lower in more compressible materials like gases.

The most compressible materials are gases. Liquids are less compressible, and solids have the least compressibility. Since air is very compressible, it would be expected to have a low velocity. Fig. 4-8 illustrates this principle. Some references call this property **stiffness**. Stiffness is the opposite of compressibility; bone, for example, has much greater stiffness than air, but is much less compressible.

To qualitatively determine the velocity of an ultrasound beam in a particular material, the effect of both density and compressibility must be evaluated. For example, if two materials have the same compressibility, the one with the greater density will have the lower propagation speed. With liquids, generally the denser the material, the lower the compressibility. Thus the loss in velocity due to higher density is about balanced by the gain in velocity due to low compressibility. This explains why most liquids transmit sound within a narrow range of velocities.

Example: Compare the velocity of ultrasound in water and mercury.

Answer: Water is about 14 times less dense than mercury. Based on the density rule, you would expect the velocity of ultrasound in mercury to be lower than in water. But water is approximately 14 times more compressible than mercury, so the velocity in mercury should be higher than in water. When considered together, the effects nearly cancel each other, and the

velocity in both liquids is nearly the same (water = 1480 m s^{-1}; mercury = 1450 m s^{-1}).

Sometimes it is readily apparent which material will transmit ultrasound with the greater velocity. For example, mercury is more compressible and more dense than bone. It is therefore not surprising that the velocity of ultrasound in bone is about three times the velocity in mercury. However, other cases are not as easy to predict. Soft tissue is more dense but less compressible than air. So one would expect the velocity in soft tissue to be faster because of compressibility and slower because of density. However, the velocity of ultrasound in soft tissue is nearly five times the velocity in air. Therefore the greater compressibility of air explains the difference in velocities.

Phase of a wave

If two waves start from the same point at different times they are said to have different phase. As shown in Fig. 4-9, locations along a sinusoidal wave can be expressed in degrees. One complete wavelength is equivalent to 360°. Therefore one half wavelength will be 180° and one quarter wavelength will be 90°.

Fig. 4-10 illustrates two waves separated by one quarter of a cycle. These waves are 90° out of phase. A shift of ½ cycle would indicate 180° phase difference. It is possible for two or more waves to occupy the same space independently of one another. If two or more waves with the same frequency have

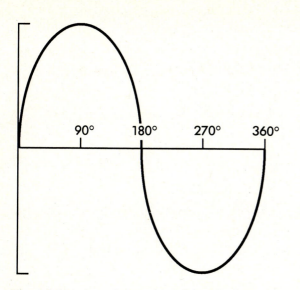

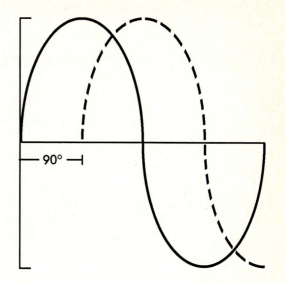

Fig. 4-9 Locations along a sinusodial wave can be expressed in degrees.

Fig. 4-10 Illustration of two waves that are 90° out of phase.

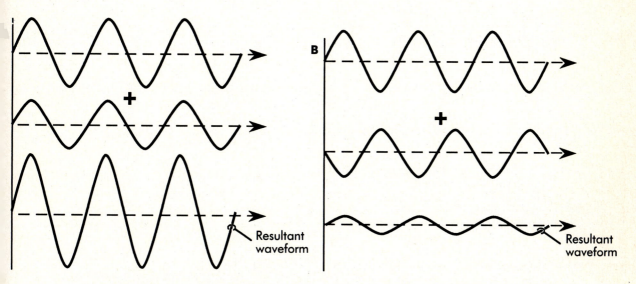

Fig. 4-11 Superposition of waves of equal frequency. **A,** Constructive interference occurs when waves of same frequency are transmittcd in phase. **B,** When waves are transmitted 180° out of phase, destructive interference results.

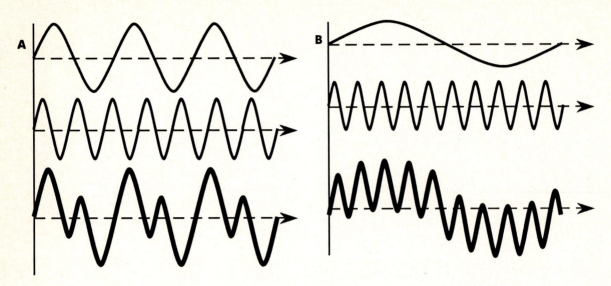

Fig. 4-12 A, The addition of two waves with different frequencies (light lines) yields complex waveforms (heavy lines). **B,** If there are large differences between frequencies, the high frequency wave appears to be superimposed upon the lower frequency.

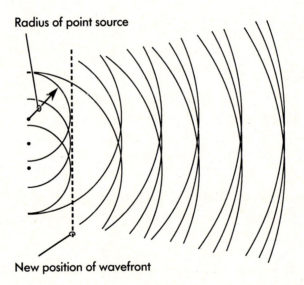

Fig. 4-13 The propagation of the ultrasound wave is described by Huygen's Principle. The ultrasound transducer behaves as though it were many individual small sources of sound, each contributing to the wave front.

the same starting points, they are said to be in phase.

Whether waves are in or out of phase, individual displacements of any particle will add together as shown in Fig. 4-11. The summation of waves to form more complex waves is known as the **principle of superposition.** Fig. 4-11, *A* shows the superposition of two waves exactly in phase that add to give a wave of the same frequency but twice the amplitude. This is an example of **constructive interference.** Fig. 4-11, *B* shows two waves that are 180°

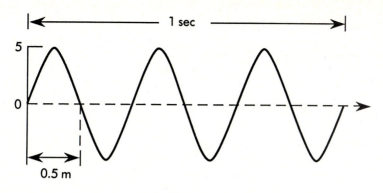

Fig. 4-14

out of phase. Every point on the top wave is exactly cancelled by the point on the lower wave. The resultant or sum is zero at all points on the superposed wave. This is an example of **destructive interference**.

Waves of different frequencies cause a complex wave that is no longer sinusoidal as when two waves of equal frequency and differing phases are added. Fig. 4-12, *A* shows the addition of two waves with the same amplitude but frequencies in the ratio of 2:1. In Fig. 4-12, *B*, the ratio of frequencies is increased to 10:1. Each component of frequencies is clearly seen in the resultant wave.

Huygen's principle

In 1678 Christen Huygen, a Dutch physicist, formulated a simple theory to explain the propagation of waves. His theory states that all points on a wave can be considered as point sources for the production of three dimensional spherical waves. Fig. 4-13 illustrates Huygen's principle in two dimensions. Three points on the initial wave front serve as centers for the secondary wavelets. The dotted line shows the position of the wave front at a later time. An ultrasound transducer, discussed in Chapter 8, can be considered a multiple point source for the generation of the ultrasonic beam.

Review Questions: Chapter 4

1. Define or otherwise identify:
  a. Transverse waves
  b. Longitudinal waves
  c. Surface waves
  d. Compressibility

  e. Phase of a wave
  f. Huygen's principle
  g. Principle of superposition

2. Compare and contrast electromagnetic waves and mechanical waves.

3. Referring to Fig. 4-14, give numerical answers with the correct units:
  a. How many wavelengths are shown?
  b. What is the wavelength?
  c. What is the frequency?
  d. What is the period?
  e. What is the amplitude?
  f. What is the wave velocity?

4. If an ultrasound beam has a frequency of 5 MHz, what is the wavelength in soft tissue in meters? In millimeters?

5. Discuss the factors, *density* and *compressibility,* that determine *ultrasound velocity.* Compare the expected effects of these factors on the velocity of ultrasound in soft tissue and bone. Which is the overriding factor for bone?

6. Matching:
  1. Period   a. reciprocal of period
  2. Velocity   b. time for a complete cycle
  3. Frequency   c. rate of soundwave travel
  4. Wavelength   d. effects velocity of propagation
  5. Density   gation
                     e. has MKS unit of "meter"

7. Discuss the factors that determine ultrasound wavelength.

8. A 4 MHz beam produces a 1 mm wavelength in a certain material.

   a. What is the ultrasound velocity in the medium?

   b. What is the velocity if the transducer frequency is changed?

   c. What frequency would be required to produce a wavelength three times as large in the same material?

5 Ultrasound Power and Intensity

As shown in Chapter 4, power measured in Watts is the rate at which work is performed. The power emitted from an ultrasound transducer is not constant; the acoustic power varies over time and space. This chapter will relate the use of power to acoustic waves and discuss other terms used to specify power and intensity of ultrasound beams.

POWER

When a 100 watt electric light bulb is turned on, 100 watts of electrical energy is converted into heat and light energy. Acoustic waves also release energy in the medium with their alternate compressions and rarefactions. However, by comparison, the amount of power made available by sound or ultrasound sources is extremely small. A full orchestra produces only 70 watts of acoustic power, less than that required by a single light bulb. The normal power output of a teacher's voice is about a million times smaller than the orchestra, except when provoked like the teacher in Fig. 5-1.

The electrical energy that jolts the piezoelectric crystal in the ultrasound transducer is converted into an ultrasound pulse that carries only milliwatts of power. Fortunately, the human ear and the ultrasound transducer are very sensitive receivers. Each can detect acoustic waves with power measured in the mW range. Power levels on a clinical ultrasound unit are generally under the control of the operator and can be changed by rotating a special knob on the unit. However, power levels are not often used to define ultrasonic output. The preferred unit, intensity, depends on the power of the beam as well as the cross sectional area of the beam.

INTENSITY

Definition: Intensity is the amount of energy per second (power) that passes through a specified area.

Symbol: I

Units: Milliwatts per square centimeter (mW cm^{-2})

Intensity can be considered as a measure of the flow rate of energy through 1 cm^2 of material positioned at right angles to the beam as seen in Fig. 5-2.

In an ultrasound beam, the more intense the beam the greater the length of vibration of particles in the medium and the greater the pressure variation found at each point in the medium as the wave passes through it. The peak velocity of particles in the medium will also increase because of the increased intensity in the ultrasound wave. Table 5-1 illustrates the effect of increasing ultrasonic intensity on particles in the medium. Focusing the ultrasound beam and increasing the power level are two ways to increase the intensity of the beam.

Amplitude

A term closely related to intensity is amplitude. Fig. 5-3 shows the representation of amplitude using a transverse wave.

Table 5-1 Acoustic variable

	Intensity	Particle displacement	Particle pressure	Particle velocity
Low intensity				
High intensity				
Peak values for 1 W cm^{-2} in water at 1 MHz		0.018 μm	1.8 atmospheres	12 cm s^{-1}

This table illustrates the effect of increasing ultrasonic intensity on three acoustic variables. A 1 W cm^{-2} intensity would be considered a high intensity beam.

Fig. 5-1 Intensity of a sound wave is related to its loudness. Students performing an experiment to see if their teacher will function as an ultrasound transducer find that only the intensity of emitted audible waves is increased.

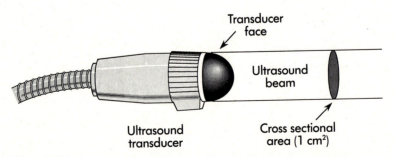

Fig. 5-2 Intensity of an ultrasound beam is power passing through a unit cross-sectional area at right angles to the direction of propagation of the ultrasound beam.

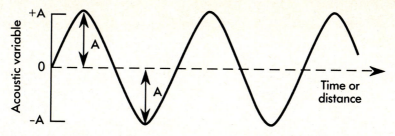

Fig. 5-3 Amplitude of an acoustic variable is the difference between the equilibrium position and the maximum value (or minimum value). It is not the difference between the maximum and minimum values. For particle displacement, the amplitude represents vibration in one direction only.

Definition:	Amplitude is the difference between the maximum value and equilibrium (average) value of an acoustic variable.
Symbol:	A
Units:	Depends on acoustic variable used: • particle displacement is in m (or μm) • particle pressure is in N m^{-2} • particle velocity is in m s^{-1}

The intensity of a wave is proportional to the amplitude of the wave squared; that is

$$I \propto A^2$$

If the amplitude is doubled, the intensity is increased fourfold. The amplitude of ultrasonic waves emitted from a transducer depends on the electrical stimulation received by the transducer. The larger the electric pulse, the more intense the beam produced and the larger the displacement disturbance transmitted to particles in a medium.

As the ultrasound beam proceeds through body tissues, the amplitude of particle vibration is reduced, thus the intensity of the beam is decreased. This loss of intensity, termed attenuation, will be discussed in Chapter 6.

The decibel

For many applications in diagnostic ultrasound, we use a unit named after Alexander Graham Bell, the inventor of the telephone. The bel is the logarithmic ratio of the relative power in two acoustic beams. The decibel (dB), $\frac{1}{10}$ of a bel, is used in sound applications to represent the smallest degree of difference in loudness that a normal human ear can distinguish. A symphony orchestra has a range of approximately 70 decibels, from a barely audible solo string to the extremely loud full orchestra. The decibel is also used to describe ultrasound attenuation and amplification.

Definition:	Decibel is a unit used to compare the relative intensities of two ultrasound beams and is expressed in logarithms to base 10.
Symbol:	Decibel (dB)
Equation:	dB = 10 Log (I/Io) (5-1)

where: I = intensity of beam at any point
Io = initial intensity of beam

Since intensity is proportional to the amplitude squared, Equation 5-1 can be written:

dB = 20 log (A/Ao)   (5-2)

Table 5-2 ranks familiar sounds according to the degree of intensity of decibel levels. The decibel values used to determine sound intensity are based

Table 5-2 Typical decibel levels for familiar sounds

Decibel level	Description of sound
130	Threshold of pain
120	Head Banger's Ball
110	Fourth of July fireworks
100	TAMU—UT football game
90	Hard Rock Cafe
80	Loud radio
70	Motorcycle ride
60	Conversational speech
50	Crackling fire
40	Typical home
30	Bambi's babies
20	Mirror Lake campsite
10	Final exam classroom
0	Ants dancing

on the ratio of the measured intensity to the standard intensity, the zero on the decibel scale.

Decibels are relative, not absolute, units; two intensities or amplitudes are required for computation of decibels. To understand this important equation, a review of logarithms will be helpful. The logarithm of a number is the power to which 10 must be raised to get that number. For example, the logarithm of 10,000 (written log 10^4) is 4; 10 must be raised to the fourth power to get 10,000. Table 5-3 shows the relationship between some powers of ten and the corresponding logarithms. Logarithms allow the compression of a wide range of numbers (for example from 1 to 10,000) into a very narrow range (0 to 4).

Example: If the transmitted ultrasound beam intensity is 1000 times greater than that reflected, what is the relative intensity?

Answer: db = 10 (log I/Io)
= 10 (log 1/1000)
= 10 (−3)
= −30 dB

In this example, the negative sign indicates a loss of intensity as the beam passes through tissue. A positive sign indicates a gain in intensity. Table 5-4 relates the relative intensity of a transmitted beam in dB to the intensity of the reflected beam and the intensity lost from the beam. To understand this table, satisfy yourself using Equation 5-1, that −10 dB results in 90% reduction in beam intensity so that only 10% of the reflected beam intensity will remain. (Hint: I/Io = 10/100 = 0.1 thus dB = 10 log (0.1).) It is essential to know that a −3 dB reduction indicates only ½ of the original beam intensity remains, thus the value **−3 dB represents a half value layer of absorber.** A loss of −6 dB means that ¼ (½ × ½) of the intensity remains.

Example: If a transducer produces an intensity of 10 mW cm^{-2} as shown in Fig. 5-4, and the returning echo is 0.001 mW cm^{-2}, what is the relative intensity?

Answer: dB = 10 log I/Io
= 10 log .001/10
= 10 log 10^{-4}
= 10 (−4)
= −40 dB

This indicates that 99.99% of the ultrasound beam is lost and only 0.01% of the original beam is present in the returning echo. If a reflected wave is returned to the transducer with 40 dB loss, the operating console of the ultrasound unit has a control to allow the sonographer to boost the intensity of the reflected wave up to +80 dB so that its size will be the same as the transmitted wave. Such an amplification gain, used to compensate for attenuation losses, will be discussed in Chapter 12.

Example: Assume intensity is reduced to ½ at a point. What is the dB loss?

Answer: dB = 10 log (.5)
dB = 10 (−.3)
dB = −3 dB Corresponds to a 50% reduction.

INTENSITY SPECIFICATION

The typical ultrasound transducer used for diagnostic applications does not deliver a uniform and continuous beam. The beam has both **spatial** (space related) and **temporal** (time related) variations.

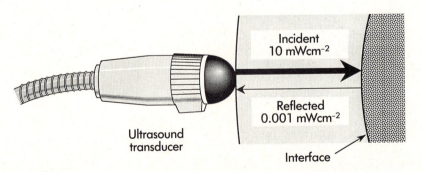

Fig. 5-4 Reflection of incident ultrasound wave at an interface.

Spatial variation

The intensity of the ultrasound beam reaches its peak at a distance from the transducer face equal to its focal length. In the focal region, the beam narrows, and the energy is concentrated into a smaller area. The beam intensity also shows variations across a cross-sectional area at some distance from the transducer. These spatial variations are represented in Fig. 5-5. The maximum intensity, known as the spatial peak (SP), is represented in Fig. 5-6. The spatial average (SA) is the average intensity across the ultrasound beam. The spatial average is generally considerably less than the spatial peak. For example, with focused transducers the spatial peak can be more than 25 times as large as the spatial average.

Temporal variation

The intensity of pulse-echo ultrasound also varies in time, since the beam is pulsed and not on continuously. The pulse consists of multiple cycles that cause intensity variations in the pulse itself. There are three measures of temporal variation, as shown in Fig. 5-7. The **temporal peak (TP)** intensity is measured at the time the ultrasound pulse is on. Averaging the intensity over one on-off beam cycle is the **temporal average (TA)**. The intensity averaged over the duration of a single pulse is called the **pulse average (PA)**.

There are six possible intensities that result from different combinations of the spatial and temporal variations of the ultrasound beam:

Table 5-3 Logarithms

Power	Number	Log₁₀
10^4	10,000	4
10^3	1,000	3
10^2	100	2
10^1	10	1
10^0	1	0
10^{-1}	0.1	-1
10^{-2}	0.01	-2
10^{-3}	0.001	-3
10^{-4}	0.0001	-4

Example: $\log 10^4 = 4$

Table 5-4 Decibels versus intensity of reflected wave

Relative intensity dB	Intensity remaining in reflected beam	Intensity lost from original beam
0 dB	100%	0%
-1 dB	79%	21%
-2 dB	63%	37%
-3 dB	50%	50%
-6 dB	25%	75%
-10 dB	10%	90%
-20 dB	1%	99%
-30 dB	0.1%	99.9%
-40 dB	0.01%	99.99%

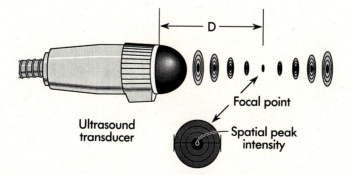

Fig. 5-5 Spatial peak occurs along central axis of beam at some distance (D), which corresponds to focal region of the transducer.

SPTP   spatial peak, temporal peak
SPPA   spatial peak, pulse average
SPTA   spatial peak, temporal average
SATP   spatial average, temporal peak
SAPA   spatial average, pulse average
SATA   spatial average, temporal average

Chapter 16 will discuss how adverse biological effects may result if excessively high ultrasound intensities are used. Consequently, the United States Food and Drug Administration requires manufac-

turers of diagnostic equipment sold in the United States to measure and report several of these intensities. Most static pulse-echo diagnostic systems operate with SPTP values in the range of 280-2800 W cm^{-2} and SPTA values that range from 20 to 1000 mW cm^{-2}. SATA intensities typically range from 10 to more than 400 mW cm^{-2}. Different types of equipment produce different intensity levels. Continuous wave Doppler units generally have larger values than the diagnostic imaging units.

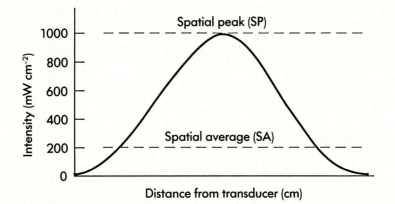

Fig. 5-6 Ultrasound beam is not uniform as it travels out from transducer. Maximum intensity of beam in space is the spatial peak (SP). Average of spatial intensity is called spatial average (SA).

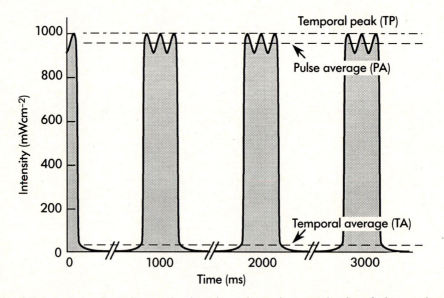

Fig. 5-7 A typical pulsed ultrasound unit emits a short microsecond pulse of ultrasound every millisecond. The temporal peak, pulse average, and temporal average intensities for a pulse-echo system are shown.

Review Questions: Chapter 5

1. Define or otherwise identify:
   a. Power
   b. Intensity
   c. Amplitude
   d. Decibel
   e. Spatial peak
   f. Spatial average
   g. Temporal peak
   h. Temporal average
   i. Pulse average

2. Discuss the concept of ultrasound intensity. What are the units? What is a typical value for a diagnostic imaging transducer?

3. How can the intensity of an ultrasound beam be increased?

4. Explain the decibel unit (dB) used in ultrasound imaging. What is the intensity lost from the beam at 1, 3, 10 and 20 dB?

5. If a beam with initial intensity of 12 mW cm^{-2} is reduced by 3 dB, what is the final intensity?

6. Distinguish between **intensity** and **amplitude**. How are they related?

7. Find the relative intensity and dB of an ultrasound beam whose transmitted intensity is 10,000 times greater than that reflected.

8. An ultrasound beam has an initial intensity of 10 mW cm^{-2}. The return echo is 0.01 mW cm^{-2}. Calculate the dB loss. What percent is lost, what percent is present in the returning echo?

9. Discuss why spatial and temporal variations occur in the ultrasound beam.

10. Of the following spatial and temporal parameters: SPTP, SPTA, SATP, SATA, SAPA,
    a. which is always the highest intensity?
    b. which is always the lowest intensity?
    c. if the ultrasound beam were continuous, which intensities would be the same?

TOMORROW WE'LL TAKE A LOOK AT THE REST OF MY SLIDE.

6 Interaction of Ultrasound with Matter

Radiographers are familiar with the interaction of x-rays and body tissues. They understand how these interactions produce an x-ray image (a radiograph) consisting of blacks, whites, and various shades of gray. Likewise, the sonographer must understand the different physical interactions that produce the ultrasound image.

It is helpful to compare x-ray and ultrasound interaction in tissues because many sonographers have a background in diagnostic radiology. The radiographic image is produced by the differential absorption of x-rays in the body. As shown in Fig. 6-1, dark areas on the radiograph are produced by x-rays that pass through the patient to the film without being absorbed. Regions of high atomic number and high density, like bone, absorb most of the incident x-rays; therefore areas beneath bone appear white on the radiograph. X-ray imaging relies on transmission of the photon beam through the body to the image receptor.

Ultrasonography, by contrast, uses the principle of **reflection.** The ultrasound image is reconstructed from the echoes reflected back from the various tissue interfaces within the body as shown in Fig. 6-2. However, as the ultrasound beam passes through tissue, the intensity of the beam is continually reduced by a process called **attenuation.** There are several different ways in which the ultrasound beam is attenuated as it passes through and reflects from body tissues. These attenuation processes include ultrasound absorption and reflection.

ATTENUATION

Definition: Attenuation is the reduction in the intensity of an ultrasound beam as it travels through a medium.

Symbol: None used-attenuation is written out

Units: Decibels (dB)

The degree of ultrasound attenuation depends on the medium involved. Therefore, an important parameter called the ultrasound **attenuation coefficient** is used to differentiate between attenuation properties of various media.

Definition: Attenuation coefficients are numerical values that express how different materials will attenuate an ultrasound beam per unit path length.

Symbol: α (Greek letter "alpha")

Units: Decibels per centimeter per MHz (dB cm^{-1} MHz^{-1})

Table 6-1 lists attenuation coefficients of common biological materials. Lung has the highest value attenuation coefficient of any biological material. This is because of the presence of small hard sacs in the lung called alveoli that scatter the ultrasound beam in all directions. Bone and air also have large attenuation coefficients. Water and blood have very low attenuation coefficients because of their low viscosity. These materials transmit ultrasound readily and are often called "ultrasound windows."

Table 6-1 Attenuation coefficients for biological materials at a frequency of 1 MHz

Material	α (dB cm^{-1})
Lung	41
Bone	20
Air	12
Soft tissue (average)	1.0
Kidney	1.0
Liver	.94
Brain	.85
Fat	.63
Blood	.18
Water	.0022

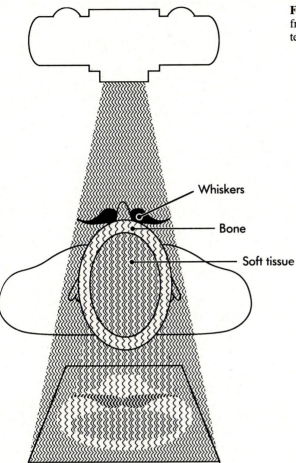

Fig. 6-1 Radiographic image density variations result from the differential absorption among tissues. Transmitted x-rays produce the radiographic image.

Whiskers

Bone

Soft tissue

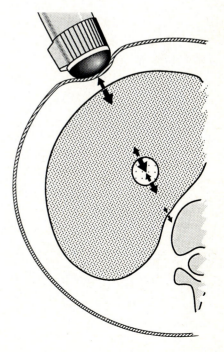

Fig. 6-2 Ultrasound image is produced by reflections from various tissue interfaces within the patient. Only those echoes that return to the transducer will contribute to the image.

Table 6-1 also shows that most soft tissues in the body, including fat, have attenuation coefficients approximately equal to 1.0 dB per cm at 1 MHz. Thus, we introduce the following **rule of thumb:**

> The attenuation coefficient in soft tissue = (6-1)
> 1 dB cm^{-1} MHz^{-1}

This rule illustrates that ultrasound attenuation depends on the frequency of operation: In tissue the rate is one dB per cm per MHz. The higher the frequency, the larger the attenuation coefficient and the less the penetration in tissue. Since the relationship between ultrasound frequency and attenuation coefficient is approximately linear, doubling the frequency doubles attenuation coefficient, tripling the frequency triples the attenuation coefficient, and so on. As discussed in Chapter 5, a decrease in ultrasound intensity is represented by a negative dB value. Thus values of α are usually preceded by a negative sign.

Example: What is the attenuation coefficient for soft tissue at 5 MHz? For air?

Answer: For soft tissue: $\alpha = (-1$ dB cm^{-1} MHz^{-1})
5 MHz $= -5$ dBcm^{-1}

For air: $\alpha = (-12$ dB cm^{-1} MHz^{-1}) 5 MHz
$= -60$ dBcm^{-1}

The attenuation coefficient can be used to compute the total attenuation if the length of the ultrasound path is known:

> Attenuation (dB) = α(dB cm^{-1} MHz^{-1}) × (6-2)
> path length (cm) × f(MHz)

Example: Find the attenuation of a 1 MHz ultrasound beam passing through 3 cm of water.

Answer: Attenuation (dB) = $(-0.0022$ dB cm^{-1}
MHz^{-1}) (3 cm) (1 MHz)
$= -.0066$ dB or only
about 1% attenuated

Example: What is the attenuation of a 4 MHz ultrasound beam after a two way trip through a 5 cm thick section of liver? (see Fig. 6-3)

Answer: Total path length of ultrasound = 2 × 5 cm
= 10 cm
Attenuation (dB) = $(-1$dB cm^{-1} MHz^{-1})
(10 cm) (4MHz)
$= -40$ dB or 99.99%
attenuation

ATTENUATION INTERACTIONS

The ultrasound transducer generates an ultrasonic beam that is directed into the patient. As the beam travels through the various tissues, its intensity is reduced. The six different interactions that are responsible for attenuation in the ultrasound beam are **absorption, refraction, diffraction, scattering, interference, and reflection.**

Each of these interactions that contribute in different amounts to the decrease of intensity of the transmitted beam will be examined in detail.

Absorption

Absorption of ultrasound is a result of internal frictional forces that oppose the vibration of molecules in tissue. The friction caused by particle movement converts the ultrasound energy into heat. Absorption is the only process that directly removes energy from the ultrasound beam. The other interactions decrease the beam intensity by redirecting the wavefront. Three factors affect the amount of absorption in the ultrasound beam.

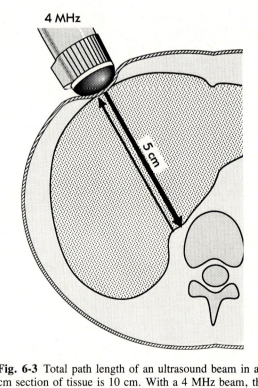

Fig. 6-3 Total path length of an ultrasound beam in a 5 cm section of tissue is 10 cm. With a 4 MHz beam, this represents attenuation of -40 dB.

Viscosity. The viscosity of a conducting medium is related to the cohesiveness or adhesiveness of the constituting molecules. High viscosity increases the internal friction of moving molecules so that energy absorption and heat production is increased. A high viscosity fluid like oil flows slowly and absorbs ultrasound more rapidly than less viscous materials like water and blood. Soft tissue is a moderate viscosity material and displays moderate ultrasound absorption. To summarize, **high viscosity results in increased absorption.**

Relaxation time. The relaxation time represents the time required for a molecule to return to its equilibrium position after it has been moved by the ultrasound wave. A short relaxation time means that a molecule quickly returns to its original position, usually before the next compression wave arrives. If the relaxation time is long, the molecules may not be able to return to their original positions before the next compression strikes them. Extra energy is required to stop and reverse the direction of such molecules. This extra energy is converted to heat and results in the increased absorption of ultrasound energy. Therefore, **long relaxation time results in increased absorption.**

Frequency. The ultrasound absorption resulting from both viscosity and relaxation time are affected by the frequency. Under the influence of high frequencies, molecules vibrate more rapidly and produce more heat in a viscous medium. With higher frequencies, successive compression waves arrive more rapidly. The molecules are not allowed as much time to relax between cycles, and more absorption results. At lower frequencies, molecules have time to relax between cycles and less absorption occurs. Therefore, **increasing frequency will increase absorption.**

Refraction

Refraction is the change in direction of ultrasound when it crosses a boundary. The familiar straw tilted in a glass of water is a good example of refraction at a water/air interface. Fig. 6-4 illustrates how the ultrasound beam is refracted as it crosses from one tissue into another.

The degree that the ultrasound beam is bent or refracted at the interface between two tissues depends on the difference in ultrasound velocity in each tissue. As the difference in velocity between tissues increases, so does the degree of refraction. The frequency of the ultrasound beam remains the same after crossing the interface, but the wavelength changes in proportion to the change in velocity. Refraction occurs only for ultrasound beams that are not perpendicular to the tissue interface. No bending of the beam occurs at perpendicular incidence.

The refraction of ultrasound obeys a principle of optics known as Snell's Law: (see Fig. 6-4).

$$\frac{\sin\theta_1}{\sin\theta_2} = \frac{v_1}{v_2} \tag{6-3}$$

where: θ_1 = incident angle
$\quad\quad\theta_2$ = transmitted angle
$\quad\quad v_1$ = velocity of sound in first medium
$\quad\quad v_2$ = velocity of sound in second medium

Sin θ is a trigonometric function that varies between 0 and 1 as θ varies between 0° and 90°.

Since soft tissue velocities of ultrasound are nearly the same, refraction does not generally produce severe problems in ultrasound imaging. However refraction can cause an object to appear to be in a

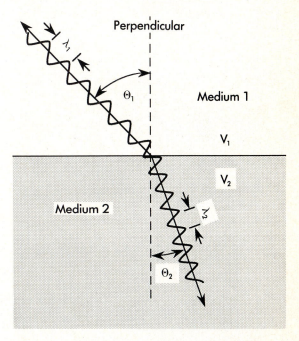

Fig. 6-4 Representation of refraction of ultrasound beam at an interface between two media that transmit ultrasound with different velocities. If the velocity is greater in medium 1, the wave is refracted toward the perpendicular and the wavelength is decreased.

different location than it actually is, or may account for the incorrect shape of an object. These are called **refraction artifacts.** An increasing degree of refraction results in increasing attenuation since the path length of the refracted beam is greater.

Scattering

Scattering is an attenuation process in which the ultrasound beam interacts with interfaces smaller than the wavelength of the beam, causing ultrasound energy to be dispersed in all directions. Fig. 6-5 depicts the scattering of an ultrasound beam by small objects such as gas bubbles, suspended particles, or red blood cells. Scattering can also be caused by

rough or irregular surfaces. This type of scatter, depicted in Fig. 6-6, is called **non-specular reflection.** **Specular reflections,** which are mirror-like reflections from smooth surfaces, provide the optimal returned signal in diagnostic imaging. This echo energy scattered back from an interface is termed **backscatter.** Non-specular reflection increases total beam attenuation.

Diffraction

Diffraction is the spreading of the ultrasound beam as it moves farther from the sound source (Fig. 6-7, *A*). The degree of diffraction is related to the size of the source. A small sound source results in large diffraction. As shown in Fig. 6-7, *B*, diffraction also occurs after the beam passes through a small opening. The aperture acts as a small source of sound, and the beam diffracts rapidly. Lateral res-

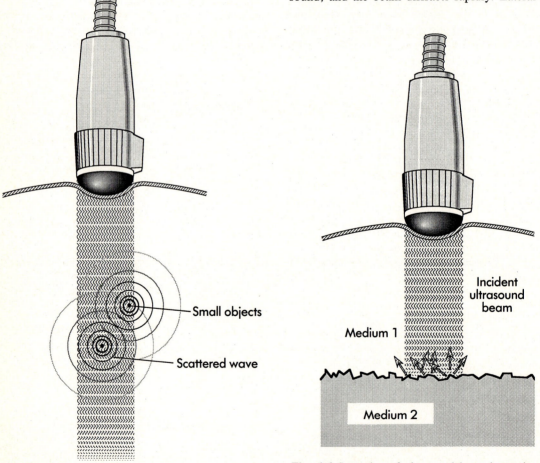

Fig. 6-5 Scattering of ultrasound beam by a small object.

Fig. 6-6 Scattering of ultrasound beam by an irregular surface.

olution, discussed in Chapter 9, is strongly affected by diffraction. **Increased diffraction results in increased attenuation.**

Interference

As discussed in Chapter 4, ultrasound waves undergo constructive interference if they are in phase. They will add together by algebraic summation to give increased amplitude. If the waves are out of phase, they undergo destructive interference. Every combination from complete constructive to complete destructive interference can occur in ultrasound. If the frequencies are similar, interference can produce the beat frequencies used in Doppler ultrasound. Wave interference is useful in transducer design since interference affects the uniformity of the beam intensity.

REFLECTION

The interaction primarily responsible for ultrasound images is reflection. If an ultrasound beam is directed perpendicular onto a large interface, it will be partially transmitted across the interface and partially reflected back toward the source as shown in Fig. 6-8. A large, smooth tissue interface, or **specular reflector,** generally produces the most useful reflection. This reflection is the echo of the pulse that returns to the transducer for use in image formation. The transmitted portion of the beam penetrates deeper into tissue and allows additional pulse echoes to be formed.

The percentage of the beam reflected at tissue interfaces depends on the beam's angle of incidence and the acoustic impedance differences of the tissues that make up the interface.

Angle of incidence

For an ultrasound beam undergoing specular reflection, the angle of incidence equals the angle of reflection as shown in Fig. 6-9. For the transducer to receive the maximum reflected signal, the sonographer must orient it so that the returning ultrasound

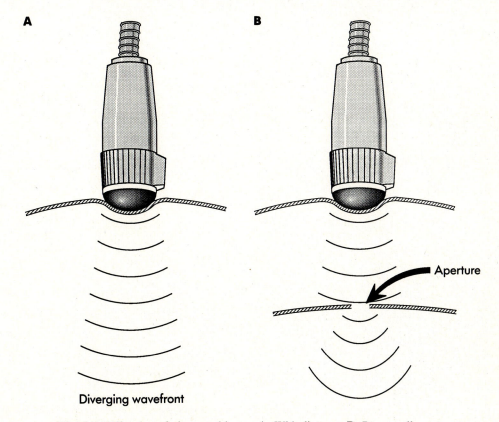

A   **B**

Aperture

Diverging wavefront

Fig. 6-7 Diffraction of ultrasound beam. **A,** With distance. **B,** By a small aperture.

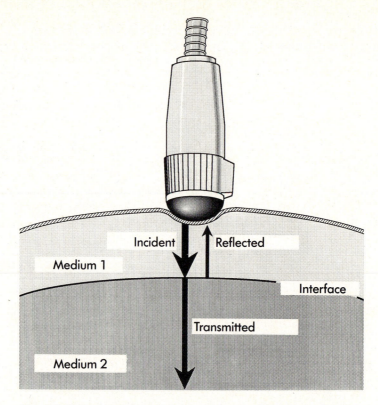

Fig. 6-8 Reflection and transmission of ultrasound beam at an interface.

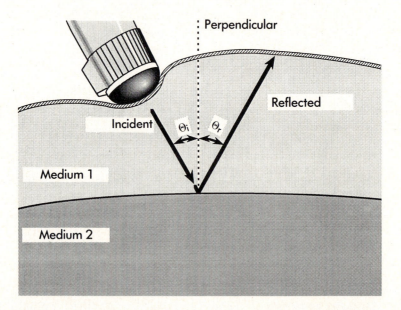

Fig. 6-9 Angle of incidence Θ_i equals the angle of reflection Θ_r for ultrasound beam reflected at an interface.

beam strikes the interface as close to perpendicular as possible. In diagnostic ultrasound, almost no reflected sound will be detected if the angle of incidence is more than 3°.

Acoustic impedance

The percentage of the ultrasound beam that is reflected at a tissue interface depends on the acoustic impedance of each tissue.

Definition: Acoustic impedance is the product of the density and the velocity of ultrasound in material. It is used in determining the amount of ultrasound reflection at an interface.

Symbol: Z

Units: Rayls

Equation: $Z = \rho v$

where ρ = density (kg m^{-3}) (Greek letter "rho")
v = velocity of ultrasound (m s^{-1})
Z = Rayls (kg m^{-2} s^{-1} × 10^{-6})

Since the velocity of ultrasound in tissue is constant over a wide range of frequencies, the acoustic impedence of a substance is constant with changing frequency. Table 6-2 lists the values of acoustic impedance for various materials. Note that the values for air (0.0004 Rayls) and bone (7.80 Rayls) are much different from those for soft tissues.

Example: Find the acoustic impedance of water if it has a density of 1000 kg m^{-3}.

Answers: $Z = \rho v$
$Z = 1000$ kg m^{-3} 1540 m s^{-1}
$Z = 1{,}540{,}000$ kg m^{-2} s^{-1}
$Z = 1.54$ Rayls

When an ultrasound beam strikes a specular reflector perpendicularly, the intensity of the reflected wave is determined by the difference in acoustic impedance between the two media that form the interface. The larger the difference in acoustic impedance at the interface, the greater the intensity of the beam reflected back toward the transducer. Conversely, if the acoustic impedance values of media on either side of the interface are the same, all of the incident beam is transmitted; none will be reflected.

To help better understand the role of acoustic impedance in reflection, consider a wave pulse set up in a length of rope as shown in Fig. 6-10, *A*. The pulse will continue down the length of rope until it

Table 6-2 Acoustic impedance for materials of diagnostic importance

Material	Acoustic impedance Rayls (kg m^{-2}s^{-1})(10^{-6})
Air	0.0004
Fat	1.38
Oil	1.43
Water	1.48
Brain	1.58
Blood	1.61
Kidney	1.62
Liver	1.65
Muscle	1.70
Bone	7.80

reaches the fastened end. If an interface is introduced by joining two ropes of very different size, the pulse is totally reflected at the interface as shown in Fig. 6-10, *B*. This situation is representative of the acoustic impedance differences and almost total reflection at a soft tissue/air interface. If the rope sizes are similar as shown in Fig. 6-10, *C*, there is partial reflection and partial transmission at the interface. Diagnostic ultrasound imaging capitalizes on the fact that the slight acoustic impedance differences at soft tissue interfaces allow both a weak reflection and strong transmission of the remaining beam to deeper structures. The reflection at the boundaries of most organs will be on the order of 1%.

The percentage of the beam reflected (%R) at an interface between two media can be computed using the intensity reflection equation:

$$\%R = \left(\frac{Z_2 - Z_1}{Z_2 + Z_1}\right)^2 \times 100 \qquad \textbf{(6-4)}$$

where Z_1 = acoustic impedance of medium 1
Z_2 = acoustic impedance of medium 2
100 is used to convert to percentage

The percent transmitted (%T) is simply the amount remaining after the %R is subtracted from 100:

$$\%T = 100 - \%R \qquad \textbf{(6-5)}$$

Note that in the formula for %R, it does not matter which of two media has the larger Z, the difference squared will eliminate a negative value.

Example: Determine the percentage of an ultrasound beam transmitted from air into soft tissue. Values of Z are taken from Table 6-2.

Answer: $\%R = \left(\dfrac{1.63 - .0004}{1.63 + .0004}\right)^2 \times 100$
$= 99.90\%$
$\%T = 100 - 99.90 = 0.10\%$

Essentially all of the ultrasound beam is reflected at a tissue/air interface; very little is transmitted to

deeper structures. Thus, for all practical purposes, imaging does not occur beyond a tissue/air interface. To avoid the complete reflection from air trapped between the transducer and the patient's skin, gel is routinely used as a coupling medium to provide good ultrasound transmission into the patient.

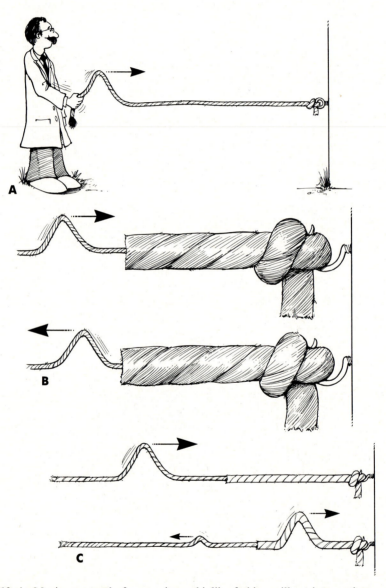

Fig. 6-10 A, Moving one end of a rope in a whiplike fashion will set into motion a wave that travels down the rope to the fastened end. **B,** Large differences in acoustic impedance result in nearly 100% reflection at the interface. **C,** If the media have similar acoustic impedances, some of the wave will be reflected, most will be transmitted.

Example: Determine the reflection and transmission at a kidney/fat interface.

Answer:
$$\%R = \left(\frac{1.62 - 1.38}{1.62 + 1.38}\right)^2 \times 100$$
$$= .64\%$$
$$\%T = 100 - .64 = 99.36\%$$

A small portion of the beam, slightly less than 1%, will be reflected back to the transducer, while the remainder is transmitted on for use in imaging deeper tissues. The thickness of either material that forms an interface does not affect the percent reflection at the interface. The strength of the reflection depends only on the acoustic impedance difference in the interface materials.

ILLUSTRATIVE EXAMPLE

The complexity of the multiple reflections and attenuation that occurs in a patient can be visualized by considering the two interfaces labeled in Fig. 6-11.

Reflection

For simplicity, assume that 100% of the beam is incident on interface A and for now ignore attenuation other than reflection. As calculated in the preceding example, of the intensity incident on interface A (fat/kidney), 99.4% is transmitted (T_1) and 0.6% is reflected back to the transducer (R_1). At interface B (kidney/air), the transmitted portion T_1 undergoes essentially 100% reflection at the air in-terface (R_2) and continues back to interface A. Here the remaining beam is divided again; 99.4% of the residual is transmitted across the interface (T_2) while 0.6% is reflected back to interface B (R_3).

This process continues (R_4, T_3, R_5) until the ultrasound beam is totally attenuated. In fact, multiple reflections (R_3, R_4, R_5 . . .) can produce troublesome reverberation artifacts that will be discussed later. Fortunately, most of the intensity remaining in these multiple reflections is quickly attenuated.

Attenuation

Recall that the rule of thumb for ultrasound attenuation in soft tissue is 1 dB cm^{-1} MHz^{-1}. For a 1 MHz transducer the incident beam (Io) and reflection R_1 will each be attenuated by -1.5 dB (-1 dB cm^{-1} × 1.5 cm) for a round trip loss within fat of -3dB (50% intensity reduction). The reflection returning from interface B that culminates in T_2 (Io + T_1 + R_2 + T_2) will have traveled a total of 11 cm (5.5 cm each way). Thus, the total soft tissue attenuation for returning pulse T_2 will be -11 dB or over 90%. Similarly, T_3 will have experienced a total path length of 19 cm that corresponds to an intensity reduction caused by attenuation of nearly 99.9%.

The multiple interfaces including crystal/tissue and soft tissue structures in the patient make the clinical situation much more complex than the simple example presented here!

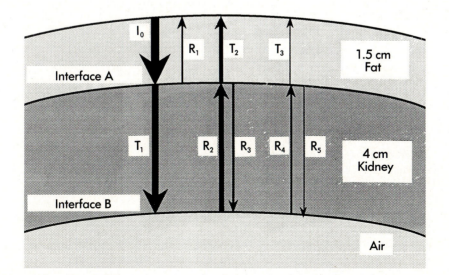

Fig. 6-11 Multiple reflection (R_i) and transmission (T_i) in an idealized three media cross-section.

Review Questions: Chapter 6

1. Define or otherwise identify:
   a. Attenuation
   b. Attenuation coefficient
   c. Absorption
   d. Refraction
   e. Scattering
   f. Snell's Law
   g. Diffraction
   h. Interference
   i. Reflection
   j. Acoustic impedance

2. Explain how ultrasound energy is absorbed from the ultrasound beam. How do viscosity, relaxation time, and frequency affect the rate of ultrasound absorption?

3. Fill a clear glass about 3/4 full of water. Insert a drinking straw or spoon, and observe it from the side of the glass. Describe what you observe. Discuss the physical phenomenon responsible for this effect, and relate it to ultrasound propagation.

4. What is the acoustic impedance of a material with a density of 10 g cm^{-3} if the velocity of sound in the medium is 2,000 m s^{-1}?

5. How would you design an ultrasound mirror? Explain your answer in terms of acoustic impedance of various materials.

6. Calculate the percent reflection and transmission for an interface where Z_1 = 4 Rayls and Z_2 = 6 Rayls.

7. Of all the soft tissues in the body which two will give the greatest percent reflection? The least percent reflection?

8. Calculate the percent reflection and transmission for a bone/fat interface. Use the acoustic impedance values listed in Table 6-2.

9. Two different materials, A and B, form an interface upon which an ultrasound beam is incident. A and B have the same thickness; A has higher Z than B. How does each of the following changes affect the percent reflection from the interface?
   a. The density of A is increased.
   b. The thickness of A is increased.
   c. The acoustic impedance of A is decreased.
   d. The frequency of the transducer is increased.
   e. The power or gain of the transducer is increased.

10. Write Snell's Law and explain each of the terms.

11. A 2 MHz ultrasound wave with an intensity of 20 mW cm^{-2} propagates through tissue. What is the intensity remaining at a depth of 5 cm? At 10 cm?

12. True or False. If false, change the italicized word to make the question true.
   a. The attenuation coefficient is largest in *bone*.
   b. *Diffusion* in ultrasound refers to the rate that sound power is absorbed per cm.
   c. A 10 cm thick slice of fat absorbs *10%* of the ultrasound beam for a 1 MHz transducer.
   d. *Scattering* can be caused by rough surfaces.
   e. 1 dB = *10 Bels*.
   f. A 3 cm section of soft tissue attenuates *more* of a 1 MHz beam than a 1 cm section of soft tissue at 5 MHz.
   g. Ultrasound absorption *decreases* as frequency increases.
   h. Ultrasound travels fastest in a *gas*.

13. Complete the following table for an ultrasound beam. *Memorize* each of these values.

Materials	Velocity (m s^{-1})	Attenuation coefficient (dB cm^{-1} MHz^{-1})	Acoustic impedance (Rayls)
Soft tissue			
Bone			
Air			
Blood			
Lung	X		X

7 The Ultrasound Transducer

Nothing has changed more dramatically in the evolution of diagnostic medical ultrasound than the size, shape, and design of the transducer. The earliest transducers were single element and cylindrical in shape; they were used for A-mode evaluations of the brain and compound B-mode imaging. More recently multielement transducers have been developed for real-time imaging using linear, phased, or annular arrays. The duplex transducer, the newest design, incorporates a single element transducer operating in the Doppler mode and a multielement array. The duplex transducer is used for anatomic and flow imaging. Each of these devices, however, involves the same physical principles to produce an ultrasound beam.

PIEZOELECTRIC EFFECT

A **transducer** is any device that converts energy from one form to another. Table 7-1 lists examples of common transducers. The ultrasound transducer converts electrical energy to mechanical energy (ultrasound) and vice versa. Similar audible sound transducers, shown in Fig. 7-1, are loudspeakers and microphones. Ultrasound transducers have no special names. They operate on the principle of piezoelectricity, literally "pressure electricity." Ultra-

sound transducers use piezoelectric materials that produce a voltage when deformed by an applied pressure. Similarly, when a voltage is applied they change shape.

The piezoelectric effect is shown schematically in Fig. 7-2. It was discovered around 1880 by the Curies of radioactivity fame. Early experiments in piezoelectricity were also conducted by Roentgen. These researchers found that piezoelectric crystals contained regions of positive and negative charge called **dipoles.** In the normal state, these dipoles are randomly arranged as shown in Fig. 7-2, *A*. If the material is heated above the **Curie Temperature** and placed in an electric field, the dipoles will align with the field and remain aligned when the crystal is cooled.

When the piezoelectric crystal is stimulated electrically, the crystal expands along its short axis. If the polarity of the electric signal is reversed, the crystal will contract. This sequence of expansion and contraction, which generates the ultrasound beam, is represented in Fig. 7-2, *B*. If the electric signal oscillates at a high frequency, the crystal will alternately expand and contract at the same frequency. In such a situation, the crystal face behaves like a high fidelity speaker cone, and this mechanical motion produces ultrasound at the same frequency as the applied electric signal (Fig. 7-3). More precisely, **an ultrasound transducer converts an electric signal into mechanical motion that results in ultrasound.**

The reverse is also possible and it too represents the piezoelectric effect. Ultrasound incident on a suitable crystalline material will transfer the energy of compression and rarefaction into contraction and expansion of the crystal. This in turn will cause an oscillating electric signal.

THE SINGLE ELEMENT TRANSDUCER

Early transducers incorporated a single piezoelectric crystal canned in a small cylindrical casing that could easily be manipulated by hand. The components of

Table 7-1 Transducers used daily

Transducer	Produces	From
Ultrasound	Mechanical energy	Electricity
Toaster	Heat	Electricity
Auto engine	Motion	Petroleum
Light bulb	Light and heat	Electricity
Generator	Electricity	Motion
Television	Light and sound	Electromagnetic radiation
Loudspeaker	Sound	Electricity
Microphone	Electricity	Sound

Fig. 7-1 The microphone (sound to electricity) and loudspeaker (electricity to sound) are examples of sonic transducers.

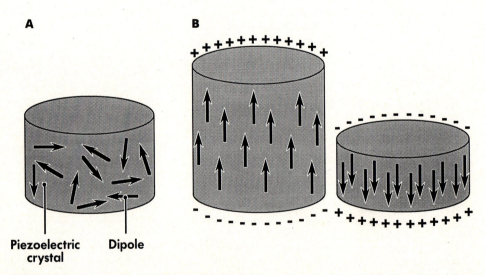

Fig. 7-2 A, Under normal conditions electric dipoles are randomly oriented. **B,** After heating, they become uniformly oriented and change orientation with an externally applied electric field.

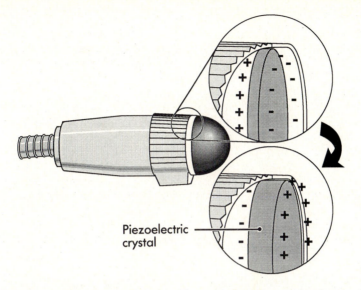

Fig. 7-3 Piezoelectric effect is the distortion of crystal shape by an electric stimulus and the generation of an electric signal by mechanical distortion of the crystal.

such a transducer, also called a **transducer assembly** or **probe,** are shown schematically in Fig. 7-4.

The case

The case is usually plastic, designed to accommodate rough handling, and yet be easily gripped and manipulated. It is normally a hermetically sealed unit that can be opened only at the manufacturer's laboratory. Electric insulation is particularly important to protect the operator from accidental shock.

The crystal

The heart of the transducer is the piezoelectric crystal or **transducer element.** Numerous crystalline materials (quartz, lithium niobate, lithium sulfate) and ceramic materials (zirconate-titanate [PZT], barium titanate, lead metaniobate) have been employed. With the exception of quartz, these crystals are not naturally piezoelectric. They are grown under very precise circumstances and conditioned by a strong electric field at high temperature so that the crystalline matrix is oriented perpendicular to the long axis. This results in a wafer-shaped object that can then have electrodes sandwiched to either surface.

There are several elastic and piezoelectric physical constants associated with PZT that makes it superior to other materials. The **strain,** measured in meters per volt (m V⁻¹), is an indication of ultrasound trans-

mitting capacity. The **generated electric field,** also measured in m V⁻¹, determines the stress-induced signal and therefore the capacity of a crystal's use as an ultrasound receiver. The **electromechanical coupling factor** indicates the efficiency of energy conversion from electrical to mechanical and vice versa. The **acoustic impedance** of the transducer element is an important characteristic because it measures the stiffness or resistance to pressure waves. PZT is the material of choice for most applications because the combination of its physical constants results in superior efficiency and higher sensitivity. Manufacturers compete for superior performance by doping PZT with impurities such as chromium, iron, nickel, and lanthanum.

The critical dimension of a piezoelectric crystal is its thickness. Each crystal has a natural resonant vibrational frequency that is related to its thickness. In order to have a single wave move back and forth between the front and back of the crystal, **the distance from one surface to the next should be one-half wavelength.** Maximum energy transfer between the mechanical and the electric state occurs when the crystal's thickness is one half the wavelength of the ultrasound or a multiple thereof. The descriptive equation that relates this property is:

$$\text{Crystal thickness} = \lambda/2 = v/2f \qquad \textbf{(7-1)}$$

where: λ = wavelength in crystal
      v = velocity in crystal
      f = frequency

For instance, a 2.5 MHz transducer produces ultrasound with a wavelength of 0.6 mm. Therefore the most effective crystal thickness would be 0.3 mm.

This thickness equal to one-half wavelength results in complete constructive interference or **resonance.** The farther from the resonant frequency one operates, the lower the efficiency of ultrasound production, and the crystal becomes less sensitive to the reception of ultrasound echoes. Typical transducer elements are 0.2 to 2 mm thick. **Thinner crystals have higher resonant frequencies.**

Example: What should the thickness be of a piezoelectric crystal designed for operation at 1.5 MHz?

Answer: $\lambda = 4000 \text{ m s}^{-1} \div 1.5 \text{ MHz}$

$\lambda = 4 \times 10^3 \text{ m s}^{-1} \div 1.5 \times 10^6 \text{ s}^{-1}$

$\lambda = 2.7 \times 10^{-3} \text{m} = 2.7 \text{ mm}$

therefore, crystal thickness $= 2.7 \text{ mm} \div 2$
                            $= 1.35 \text{ mm}$

For such a calculation, one must use the velocity of ultrasound in the crystal (4000 m s^{-1}, not the velocity in soft tissue).

Each surface of the piezoelectric crystal is metal coated to make it electrically conductive. Electrodes are fastened to each surface and then attached to an electrical connector at the inactive end of the trans-

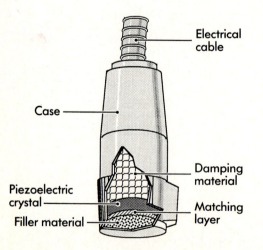

Case

Piezoelectric crystal

Filler material

Electrical cable

Damping material

Matching layer

Fig. 7-4 Components of a typical single element transducer.

ducer assembly. Heat in excess of 350°C will destroy the piezoelectric property of the crystal by altering the alignment of the dipoles. **Never autoclave a transducer.**

Damping material

Immediately behind the piezoelectric crystal is **damping material,** usually a combination of an epoxy resin and tungsten powder, designed to damp the ultrasound pulse. This damping material behaves much like the member of the orchestra seen in Fig. 7-5 who bangs two cymbals together and then damps their noise by pressing them to his chest.

Ideally the damping material, like the cymbalist's body, absorbs most of the ultrasound emitted by terminating the "ringing" of the crystal. Damping reduces the sensitivity of the transducer because it lowers the intensity of the output signal. However, damping improves axial resolution by typically reducing the number of cycles in each pulse to three to five as shown in Fig. 7-6. This effectively shortens the pulse and reduces the pulse duration.

In a diagnostic ultrasound transducer the damping material must have an acoustic impedance comparable to that of the crystal in order to get maximum absorption and to ensure a short duration pulse. Conversely, in therapeutic ultrasound it is necessary to deliver the maximum amount of energy into the patient. Therefore transducers designed for therapy use air for damping material since the acoustic impedance mismatch at the crystal/air interface will ensure that nearly 100% of the beam is reflected into the patient.

Matching layer

A layer of material with an acoustic impedance between that of the piezoelectric crystal (Z = 30 rayls) and soft tissue (Z$_{avg}$ = 1.6 rayls) is sandwiched to the face of the transducer. Use of such intermediate acoustic-impedance material allows more of the ultrasound beam to be transmitted into the patient because less of the beam is reflected back to the transducer. This matching layer, like the damping material, is usually composed of an epoxy resin-tungsten powder matrix designed to reduce the reflection of the ultrasound emission at the crystal's surface.

The matching layer is usually designed to be one quarter of a wavelength thick. Such a design is called the **quarter wavelength matching layer transducer.** Since a diagnostic ultrasound pulse contains multiple frequencies and hence, multiple wave-

Fig. 7-5 The cymbalist stands in the back of the orchestra, clangs the cymbals together, and clamps the noise against his chest.

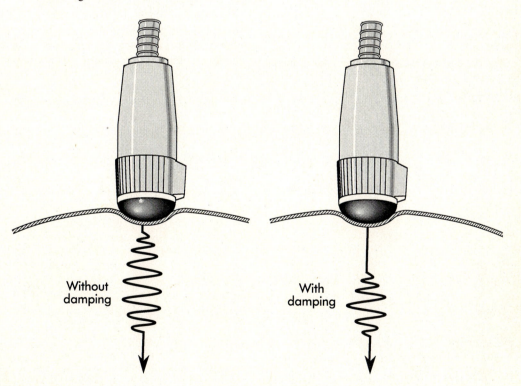

Without damping

With damping

Fig. 7-6 Damping the ultrasound beam shortens pulse length, reduces pulse duration and improves axial resolution.

lengths, the quarter-layer thickness is normally determined from the center frequency of the transducer. Some transducers use multiple matching layers to further improve the transfer of the ultrasound energy into tissue.

Air has a very low acoustic impedance. Any air trapped between the transducer face and the skin will result in nearly total reflection. Little if any of the ultrasound beam will be transmitted into soft tissue. Therefore, one must use a copious amount of gel or mineral oil to produce a good **acoustic coupling** with the skin and permit the ultrasound to pass into soft tissue.

QUALITY FACTOR

The more cycles present in an ultrasound pulse, the more one can be sure of the purity of the frequency of that ultrasound emission. Continuous wave ultrasound has essentially one frequency, the **resonant frequency** of operation seen in Fig. 7-7, *A*. Relatively little time is spent starting and stopping a continuous wave ultrasound beam.

Pulsed ultrasound, on the other hand, may contain many frequencies in each pulse because of engineering difficulties. It is not possible to instantaneously start or stop an ultrasound pulse. It takes some time, though short, to reach the resonant fre-

quency and an equal time to terminate such a pulse. An ultrasound pulse may require 50% of its time starting and stopping as seen in Fig. 7-7, *B*. In general the fewer cycles there are in an ultrasound pulse the more frequencies it will contain. Conversely, the longer an ultrasound emission, the closer it will be to constant frequency.

Sonographers have adopted an electrical engineering term, **bandwidth,** to express the range of frequencies in an ultrasound beam. For example, a pure continuous wave emission at 2.25 MHz would have a bandwidth of zero. A typical pulsed emission would contain frequencies ranging from 1.75 to 3 MHz because of difficulties starting and stopping the pulse. Its bandwidth therefore would be greater, as shown in Fig. 7-7, *B*. Bandwidth is determined by transducer fabrication and design of the electronics.

Another borrowed electrical engineering term is Q-value or **quality factor** (QF). The quality factor describes the purity of vibration of the piezoelectric crystal. It also describes the frequency homogeneity of the ultrasound beam.

A high Q transducer "rings" for a long period of time producing a long pulse length. A long "ring down time" or high Q transducer is good for therapy or continuous wave diagnostic ultrasound. Low Q transducers have short "ring down time" and there-

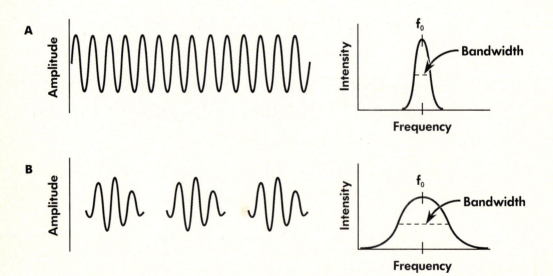

Fig. 7-7 A, Continuous wave ultrasound has essentially only one frequency, f_o. **B,** Pulsed ultrasound has a spectrum of frequencies.

fore short pulse length because fewer cycles are contained in the pulse. They are most useful for diagnostic imaging. High and low quality ultrasound beams are shown schematically in Fig. 7-8.

Quality factor is determined by dividing the resonant frequency of the ultrasound beam by its bandwidth. This is described mathematically in Equation 7-2.

Quality Factor = (resonant frequency) ÷
(bandwidth) **(7-2)**

The narrower the bandwidth, the higher the quality factor. As a rule of thumb, the number of cycles in a short ultrasound pulse will be approximately equal to the Q-factor. Therefore, a five-cycle pulse emission will have a quality factor of five. A continuous emission will have a much higher quality factor.

Example: What is the Q factor of a 2.25 MHz transducer with a bandwidth of 1.25 MHz?

Answer: QF = (2.25 MHz) ÷ (1.25 MHz) = 1.8

Table 7-2 is a summary of the properties associated with high and low Q transducers.

Table 7-2 Transducer performance as it relates to Q value

High Q transducers	Low Q transducers
Have long ring down time	Have short ring down time
Have narrow range of frequencies	Have wide frequency range
Make better transmitters	Make better receivers
Are used for therapy and CW Doppler	Are used for pulse-echo imaging

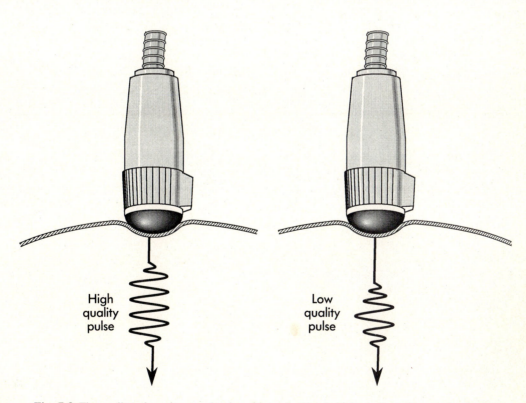

Fig. 7-8 The quality of an ultrasound pulse of long duration is high. A short ultrasound pulse has low quality because of the shorter time spent at the resonant frequency.

Review Questions: Chapter 7

1. Define or otherwise identify:
   a. Piezoelectricity
   b. Curie temperature
   c. Transducer
   d. Damping material
   e. Quality factor

2. An ultrasound transducer converts electrical energy to ultrasound and ultrasound to electrical energy. What transducer devices accomplish the same for sonic waves?

3. Explain with diagrams how a piezoelectric crystal converts electric energy into ultrasound.

4. What are the steps required to produce a piezoelectric crystal?

5. List three materials that are used as the piezoelectric crystal in an ultrasound transducer.

6. What is the unit of measure for the elastic constant associated with a piezoelectric crystal?

7. Which type of crystal is usually used in diagnostic ultrasound transducers? Why?

8. What is the critical dimension of a piezoelectric crystal? Why?

9. What should be the thickness of the piezoelectric crystal designed for operation at 5 MHz?

10. Describe the most desirable value of acoustic impedance and thickness for the matching layer of an ultrasound transducer.

11. One PZT crystal is 0.2 mm thicker than another. If they are each used in a pulsed transducer, what will be the difference in frequencies between the two transducers?

WE'LL BEGIN OUR TESTS WITH A SONOGRAM OF YOUR WALLET.

8 The Ultrasound Beam

Earlier, in Chapters 4 and 5, diagnostic ultrasound was characterized as a continuous mechanical wave with alternating regions of compression and rarefaction propagating through a conducting medium, normally soft tissue. In actual clinical imaging application, intermittent pulses of ultrasound, each containing several regions of compression and rarefaction called **cycles,** radiate from the transducer face. Regardless of whether the beam is emitted as a continuous wave or as a pulsed wave, it will have certain macroscopic characteristics of importance to the sonographer. The characteristics of pulse wave ultrasound are of particular importance to diagnostic imaging.

CHARACTERISTICS OF ULTRASOUND PULSES

The terms **wavelength, frequency,** and **period,** introduced in Chapter 4, were shown to describe characteristics of a continuous wave. However, in most diagnostic ultrasonography a pulse-echo technique is used in which a pulse is emitted and ultrasound echoes are received by the transducer, after some time delay. The transducer emits a pulse by vibrating for three to five cycles and then gathers information from returning echoes before emitting the next pulse. Fig. 8-1 demonstrates how the terms used to describe continuous ultrasound beams also apply to the cycles within an ultrasound pulse. There are several additional terms that are exclusively applied to pulse-echo ultrasound.

Pulse repetition frequency

Definition:	Pulse repetition frequency is the pulse rate or the number of pulses emitted per second. (Remember that each pulse contains several cycles.)
Symbol:	PRF
Units:	Pulses/second or Hertz (Hz)

A pulse repetition frequency of 4 pulses per second or 4 Hz is shown in Fig. 8-2. Some ultrasound equipment manufacturers use a pulse repetition frequency of 1000 pulses per second or 1 kHz. Others automatically vary the PRF with depth. For real-time imaging, it is advantageous to pulse as rapidly as possible so the transducer can be moved without skipping a portion of the anatomy. However, since ultrasound travels at a fixed speed in tissue, pulses must be spaced far enough apart in time so that all distant echoes will have enough time to return to the transducer. Thus the maximum pulse repetition frequency of an instrument is limited to about 2500 Hz.

Pulse repetition period

Definition:	Pulse repetition period is the time from the beginning of one pulse to the beginning of the next.
Symbol:	PRP
Units:	Seconds (s)

The pulse repetition period consists of the time the transducer is actively producing a pulse plus the time that it listens for return echoes. Fig. 8-3 illustrates this principle with a 1000 μs (1m s) PRP. The pulse repetition period and the pulse repetition frequency are reciprocally related.

Equation:	PRP = 1/PRF	(8-1)

Example:	Find the PRP if the PRF is (a) 2 Hz (b) 1000 Hz?
Answer:	a. PRP = $1/2s^{-1}$ = 0.5s
	b. PRP = $1/1000\ s^{-1}$ = 0.001s = 1 ms

Pulse duration

Definition:	Pulse duration is the time during which the pulse actually occurs.
Symbol:	PD

Pulsed wave

Wavelength = 0.3 mm, Frequency = 5 MHz, Period = 0.2 μs

Continuous wave

Fig. 8-1 Wavelength, frequency, and period apply to ultrasound pulses as well as continuous wave ultrasound, shown here for 5 MHz in soft tissue.

|—— 250 ms ——|

|——————————— 1 s ———————————|

Fig. 8-2 These ultrasound pulses exhibit a pulse repetition frequency of 4 Hz. Note that each pulse contains 4 cycles.

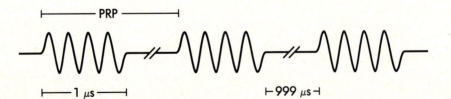

|——— PRP ———|

|— 1 μs —| |— 999 μs —|

Fig. 8-3 The pulse repetition period is the time from the beginning of one pulse to the beginning of the next pulse, in this case 1000 μs or 1 m s.

Units: Seconds (s)

The pulse duration is the pulse "on" time. It does not include the time the transducer is silent and no ultrasound is being emitted. Pulse duration is the product of the period (not the pulse repetition period) and the number of cycles in a pulse.

PD = period × number of cycles in a pulse	(8-2)

Example: If a 2 MHz transducer has 3 cycles in a pulse, find the pulse duration.

Answer: Period = $1/f = 1/(2 \times 10^6 \text{ s}^{-1})$
            = 0.5×10^{-6} s
         PD = $(0.5 \times 10^{-6} \text{ s})$ 3
         PD = 1.5×10^{-6} s = 1.5 μs

Duty factor

Definition:	Duty factor is the fraction of time the ultrasound pulse is actually being emitted during one pulse repetition period.
Symbol:	DF
Units:	None

The duty factor is a dimensionless ratio that expresses the fraction of time that the transducer is actually emitting ultrasound. Therefore, **for continuous wave ultrasound the duty factor is one;** for pulse echo ultrasound the duty factor is very low. The duty factor is mathematically equal to the quotient of the pulse duration divided by the pulse repetition frequency.

DF = pulse duration/pulse repetition period	(8-3)

Example: What is the duty factor of an ultrasound beam with a pulse duration of 1.5 μs and a pulse repetition period of 1 ms?

Answer: DF = 1.5 μs/1ms = 1.5×10^{-6} s/1 × 10^{-3} s = 1.5×10^{-3}
         DF = .0015

This result indicates that the transducer emits ultrasound only 0.15% of the time. The remainder of the time it listens for returning echoes.

Spatial pulse length

Definition:	Spatial pulse length (SPL) is the length or space over which an ultrasound pulse occurs.
Symbol:	SPL
Units:	m

One tends to think of an ultrasound pulse existing for a certain length of time, the pulse duration. How-ever, the important characteristic for imaging is the length of the pulse in tissue or the spatial pulse length; the SPL affects image resolution greatly. Ultrasound pulses with long SPL will interact with a tissue interface longer than pulses with a short SPL, which will result in a blurred image of the interface.

The spatial pulse length is defined as the product of the ultrasound wavelength multiplied by the number of cycles in a pulse.

SPL = λ × number of cycles in pulse	(8-4)

Example: A 2 MHz transducer produces a 3-cycle pulse. Find the SPL in soft tissue.

Answer: λ = v/f = 1540 m s^{-1}/2 × 10^6 s^{-1}
            = 7.7×10^{-4} m
         λ = 0.77 mm
       SPL = (0.77 mm) 3 cycles = 2.3 mm

It is important not to confuse the wave properties **wavelength, frequency,** and **period** that are characteristics of the ultrasound beam with the pulsing characteristics described above.

THE WAVE FRONT

Whether continuous wave or pulsed, diagnostic ultrasound is emitted from the face of a transducer in wave-like fashion. As discussed in Chapter 4, ultrasound waves are not unlike ocean waves at the seashore. An even closer analogy would be high fidelity audible sound. Consider the three audio speakers in Fig. 8-4. In general, as the source of sound becomes smaller, the emission of sound becomes more isotropic. **Isotropic** means with equal intensity in all directions.

If the smallest speaker in Fig. 8-4 had a diameter of ½ wave length (λ/2) the emitted sound pattern would appear as in Fig. 8-5. Here the sound emission folds back on itself creating a **spherical wave front** emitted from the transducer face. If the transducer diameter were even smaller, the emitted beam would be more spherical in shape and more isotropic in intensity.

If this single transducer of Fig. 8-5 were replaced by a line of several similar small transducer elements each emitting sound isotropically, the phenomenon of constructive and destructive interference would result. Alternate regions of enhancement and reduction of the sound wave would appear as it emerged from these multiple sources. Such a multiple source generator of ultrasound creates a **plane wave front** as shown in Fig. 8-6. This same phenomenon occurs with a properly designed very large transducer. It is this plane wave front that is employed for imaging

Fig. 8-4 For any given frequency, the larger the sound source, the more directional will be the sound.

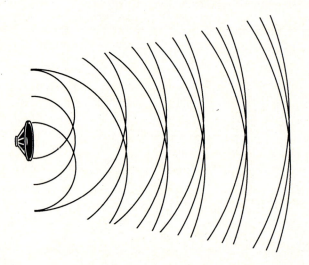

Fig. 8-5 Spherical wave fronts are emitted from a transducer with a one-half wavelength diameter.

with diagnostic ultrasound, whether from a single large transducer or an array of small transducer elements.

The large transducer behaves as though it were many individual small sources of sound, each contributing to the wave front. The transducer array is, in fact, a collection of small transducer elements. These interference patterns produce the plane wave front. A single large transducer can be focused, much like a camera lens. The transducer array can be focused and steered, and this represents a significant advantage of the transducer array over the

single element transducer. Focusing and steering are covered later.

NEAR FIELD AND FAR FIELD

A transducer whose single crystal diameter is an intregral multiple of the primary wavelength will emit an ultrasound beam characterized by a strong plane wave front. Likewise, a multiple transducer array will emit a strong plane wave front. This plane wave front has two distinct regions, each with different physical characteristics. These features of the ultrasound beam are shown schematically in Fig. 8-7.

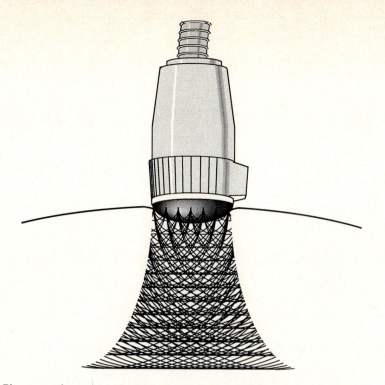

Fig. 8-6 Plane wave fronts are emitted from large single transducers and multiple transducer arrays.

Near field   Far field

$$\frac{d^2}{4\lambda}$$

$$\text{sine } \Theta = \frac{1.22\lambda}{d}$$

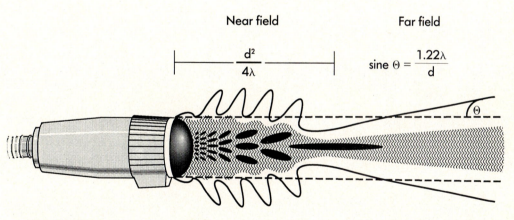

Fig. 8-7 There are two distinct regions to the ultrasound beam, the near field and the far field.

The description of these two ultrasound beam regions is borrowed from optical physics. The region nearest the transducer face is called the **near field** or **Fresnel Zone,** and it is characterized by a highly collimated beam with great variation in ultrasound intensity from wave front to wave front.

The region farthest from the transducer face is called the **far field** or **Fraunhofer Zone.** The far field is characterized by a divergence of the ultrasound beam and a more uniform ultrasound intensity

from wave front to wave front. **Best imaging is obtained at the near field, far field transition.** This is also the natural focal length of a flat-faced transducer. Some ultrasound side lobes exist outside the primary emission and, although these are relatively unimportant, they may interfere with image quality if not properly handled.

For a single element transducer, the diameter and length of the near field and the divergence of the far field are determined by the transducer diameter and

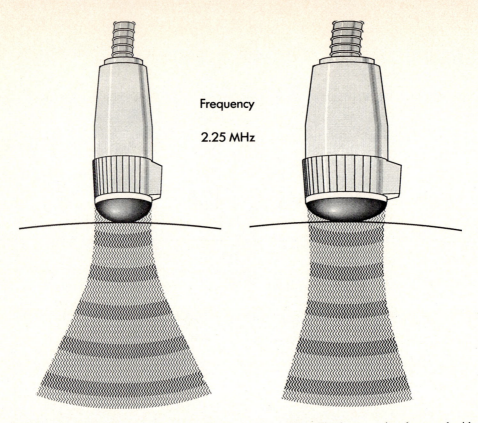

Frequency

2.25 MHz

Fig. 8-8 Both of these transducers are operating at 2.25 MHz. The larger emits ultrasound with a longer near field and a more collimated far field.

the ultrasound frequency. The **length of the near field** (L) in soft tissue is given by the following equations:

$$L = r^2 \div \lambda \qquad (8\text{-}5)$$

$$L = d^2 \div 4\lambda \qquad (8\text{-}6)$$

$$L = d^2 f \div 4v \qquad (8\text{-}7)$$

where r = transducer radius
d = transducer diameter
λ = ultrasound wavelength
f = ultrasound frequency
v = ultrasound velocity

Since wavelength and frequency are inversely related, either may be used to determine the length of the near field.

Example: A 12 mm diameter transducer emits ultrasound with 0.5 mm wavelength in soft tissue. What is the length of the near field?

Answer: L = $(6 \times 10^{-3}$ m$)^2 \div 0.5 \times 10^{-3}$m
L = 36×10^{-6} m$^2 \div 0.5 \times 10^{-3}$ m
L = 72×10^{-3} m
L = 72 mm

Example: What is the length of the near field for a 20 mm diameter transducer that emits ultrasound with a 0.8 mm wavelength in soft tissue?

Answer: L = $(20 \times 10^{-3}$ m$)^2 \div (4)(0.8 \times 10^{-3}$ m$)$
L = 400×10^{-6} m$^2 \div 3.2 \times 10^{-3}$ m
L = 125×10^{-3} m
L = 125 mm

Example: Compute the wavelength of 5 MHz ultrasound, and use this value to determine the focal length of a 15 mm diameter transducer.

Answer: λ = $(1540$ m s$^{-1}) \div (5 \times 10^6$ s$^{-1})$
= 308×10^{-6} m = 0.3 mm
L = $(15 \times 10^{-3}$ m$)^2 \div 4 (0.3 \times 10^{-3}$ m$)$
L = 225×10^{-6} m $\div 1.2 \times 10^{-3}$ m
L = 187.5×10^{-3} m
L = 188 mm

Ultrasound emission beyond the length of the near field is said to be in the far field. The far field is characterized by a diverging beam having less intensity variation from wave front to wave front. The divergence of the far field is expressed by the angle

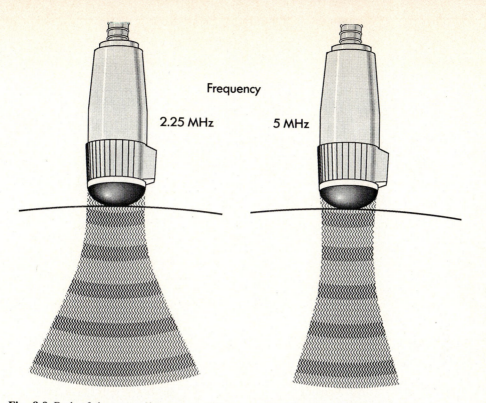

Frequency

2.25 MHz   5 MHz

Fig. 8-9 Both of these transducers are of the same size. The one operating at higher frequency emits ultrasound with a longer near field and a more collimated far field.

of divergence θ with the central axis of the beam and can be computed as follows.

$$\text{Sin } \theta = (1.22\lambda) \div d \qquad (8\text{-}8))$$

Example: What is the divergence of ultrasound in the far field from a 20 mm transducer operated at 2.25 MHz?

Answer: at 2.25 MHz the wavelength is:

$$\lambda = (1540 \text{ m s}^{-1}) \div 2.25 \times 10^6 \text{ s}^{-1}$$
$$\lambda = 684 \times 10^{-6} \text{ m}$$
$$\lambda = 0.7 \text{ mm}$$

therefore:

$$\text{Sin } \theta = (1.22 \times 0.7 \text{ mm}) \div 20 \text{ mm}$$
$$\text{Sin } \theta = 0.043$$
$$\theta = 2.45°$$

The previous discussion is not precisely correct. For a flat disc-type transducer there is actually some narrowing of the ultrasound beam. Minimum beam size actually exists at the near field-far field transition. However, this is relatively unimportant since most such transducers are mechanically focused anyway.

In general one can make the following statements relating transducer design to ultrasonic beam properties:

1. As the transducer diameter is increased, with constant frequency, the near field is lengthened and the far field does not diverge as much (Fig. 8-8).
2. As the ultrasound frequency is increased, with fixed diameter, the near field is lengthened and the far field does not diverge as much (Fig. 8-9).

Table 8-1 shows some representative field properties of commercially available non-focused transducers.

Since most ultrasound imaging today is conducted with multielement transducer arrays, we may consider that these same relationships hold. However, there are subtle differences when we take full advantage of the ability to manipulate the ultrasound beam from an array through steering and focusing.

FOCUSING THE ULTRASOUND BEAM

With a single element disc-type transducer the beam width in the imaging region is a function of the transducer diameter. The smaller the transducer diameter is, the better the lateral resolution. Lateral resolution is fully discussed in Chapter 9.

The lateral resolution can also be improved by shaping the face of the transducer as illustrated in

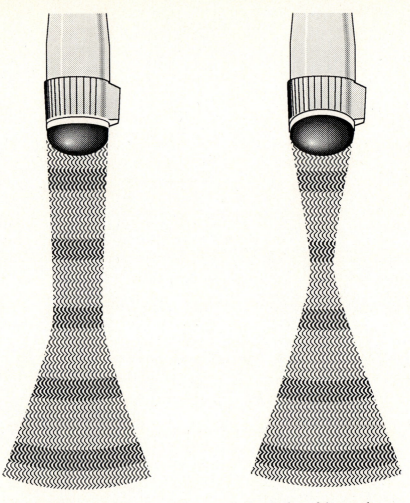

Fig. 8-10 Focusing can be accomplished by shaping the face of the transducer.

Fig. 8-10. Having a concave transducer face reduces the plane wave front in diameter at the near field–far field transition, and this improves the lateral resolution. As the transducer face is made more concave, the beam becomes even more focused.

High frequency transducers cannot be focused by shaping because the piezoelectric crystal is too thin. Such transducers must be focussed with plastic lenses as shown in Fig. 8-11. These lenses can be placed at any point in the ultrasound beam, but are normally attached to the face of the transducer. Acoustic lenses take advantage of the property of sound refraction that changes the direction of the ultrasound wave. Such lenses are made of polystyrene, nylon, other plastic materials, and aluminum.

Table 8-1 Near field lengths and far field divergence of nonfocused transducers

Transducer diameter (mm)	Frequency (MHz)	Near field length (cm)	Far field divergence (degrees)
6	10	6.0	1.79°
8	10	10.7	1.35°
8	5	5.3	2.69°
10	5	8.3	2.15°
10	2.5	4.2	4.31°
15	2.5	9.4	2.87°
15	1.0	3.8	7.20°
20	1.0	6.7	5.39°

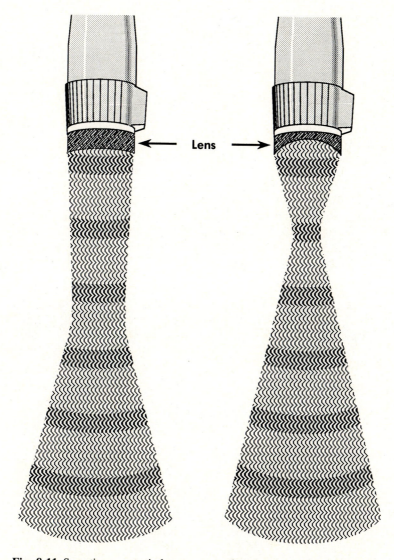

Fig. 8-11 Sometimes acoustic lenses are employed to focus ultrasound beams.

Review Questions: Chapter 8

1. Define or otherwise identify:
   a. Pulse repetition frequency
   b. Pulse repetition period
   c. Pulse duration
   d. Duty factor
   e. Spatial pulse length
   f. Isotropic

2. An ultrasound unit produces 2 ms pulses, 48 ms apart.
   a. Draw the pulse sequence showing two pulses and the beam-off time between them.
   b. What is the pulse repetition frequency?
   c. What is the pulse repetition period?
   d. What is the pulse duration?
   e. What is the duty factor?

3. A 4 MHz transducer emits 4 cycles in a pulse and has a pulse repetition frequency of 500 Hz.
   a. What is the pulse repetition period?
   b. What is the pulse duration?
   c. What is the duty factor?
   d. What is the spatial pulse length in soft tissue?

4. What is the length of the near field of an 18 mm diameter transducer having a 0.6 mm wavelength in soft tissue?

5. What is the focal length of a 24 mm diameter transducer operated at 2.25 MHz?

6. A 15 mm diameter transducer is operated at 7.5 MHz. What is the angular divergence in the far field?

7. Discuss two methods of focusing transducers.

8. What is the effect of increasing the diameter of an ultrasound transducer (fixed frequency)?
   a. Length of the near field
   b. Beam width in the near field
   c. Wavelength
   d. Beam width in the far field

9. *True or False*
   a. Focusing changes the focal length of the beam. _____
   b. As transducer diameter increases, beam diameter in the far field increases. _____
   c. Focusing can be accomplished with a plastic lens. _____
   d. The far field is known as the Fresned zone. _____
   e. Near field length can be increased by increasing transducer frequency. _____
   f. The far field divergence angle decreases as the transducer diameter is increased. _____

I WOULDN'T WORRY ABOUT IT, THERE'S NOTHING YOU CAN DO. THE DIVERGENCE OF THE FRAUNHOFER ZONE IS ALWAYS SINE Θ DEPENDENT!

BARTENDER WITH PH.D. IN ULTRASOUND PHYSICS

9 Ultrasound Resolution

When characterizing radiographic images, radiologic technologists classically employ terms such as detail, definition, and distortion. **Detail** relates to the ability of the radiographic image to faithfully reproduce small objects. **Definition** refers to faithfully reproducing larger objects that have little tissue differentiation. **Distortion** results from the unequal magnification of anatomical objects that lie different distances from the image receptor.

These classical characterizations of radiographic images are used less because of the digital imaging techniques now available: computed tomography (CT), magnetic resonance imaging (MRI), and digital sonography. For all such images there are again three principal characteristics, but they differ from those of radiographic imaging. **Spatial resolution** is somewhat analogous to detail in that it measures the ability of an imaging system to faithfully reproduce very small, high contrast objects. **Spatial resolution refers to the size of the smallest high contrast object that can be imaged.** Radiographic imaging can reproduce objects as small as 0.1 mm, while CT and MRI can image high-contrast objects only as small as 0.5 mm. Spatial resolution in diagnostic ultrasound is about 1 mm.

The second principal measure of digital image quality is **contrast resolution or the ability to differentiate anatomic structures having similar tissue characteristics.** Contrast resolution is generally expressed in units of size at a given contrast level. For instance, a good CT scanner can visualize a 4 mm object separated from surrounding tissue with a contrast of 0.4%. That's a difference of 4 Houndsfield Units between the two tissues. MRI, on the other hand, can image 2 mm at 0.1% contrast with ease. Similar terms are not used to describe diagnostic ultrasound because its contrast resolution is primarily a function of interface characteristics and not tissue attenuation.

These factors of spatial and contrast resolution are often combined into a contrast-detail curve such as that shown in Fig. 9-1. This representative relationship shows that both CT and MRI have approximately the same spatial resolution, while radiography provides the best spatial resolution. Contrast resolution, on the other hand, is superior with MRI because of the intrinsic properties of the magnetic resonance parameters: spin density, T1 relaxation, and T2 relaxation. Ultrasonic contrast resolution is not as good as that of CT, but it is better than that available with radiography. Both the spatial resolution and the contrast resolution of diagnostic ultrasound is worse than CT and MRI, but it is also less expensive and totally adequate for most of its applications.

Noise is the third principle characteristic of a medical image. Noise can affect the spatial resolution of a system, but it has much more effect on contrast resolution. The presence of noise obscures the definition of low contrast objects. Contrast resolution for most systems is limited by the system's noise, and ultrasonic noise is relatively high.

Ultrasound images are obtained from reflections at interfaces, and have what could be described as good contrast resolution. The contrast, however, is due to structural differences at interfaces and to subtle differences in tissue composition. Some research work continues in attempts to characterize various tissues with ultrasound, but to date this has been clinically unproductive.

Spatial resolution is the single image characteristic that is employed to characterize the ultrasonic image. When internal structures are scanned with an ultrasound beam to produce a two-dimensional image, the quality of the image will be directly related to the spatial resolution of the system. Spatial resolution in diagnostic ultrasound is the ability of the system to identify closely separated tissue interfaces

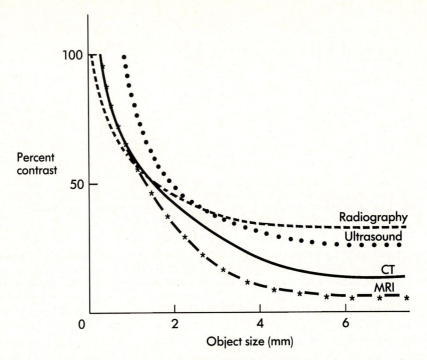

Fig. 9-1 A contrast detail curve relates considerable information about the spatial and contrast resolution of an imaging system.

lying along the axis of the beam or perpendicular to the axis of the beam. Therefore, proper specification of ultrasound image quality involves **axial resolution** and **lateral resolution.**

AXIAL RESOLUTION

Definition: Axial resolution is the ability to image closely spaced interfaces on the axis of the ultrasound beam.

Symbol: AR

Units: Millimeters (mm)

Axial resolution sometimes called **range, depth, or longitudinal resolution** is a measure of the ability of the ultrasound system to identify closely separated interfaces that lie on the axis of the ultrasound beam. Fig. 9-2 illustrates this property. If the interfaces are widely separated they will be easily identified. As the interfaces become closer together, however, the returning echoes from each interface may not be distinguishable from each other, and separately returning echoes appear as one.

Axial resolution depends on essentially three characteristics of the ultrasound beam:

1. The spatial pulse length
2. The ultrasound frequency
3. The damping factor

These factors are interrelated of course.

Spatial pulse length

Axial resolution depends primarily on the length of the ultrasound pulse in space. Therefore, a pulse of ultrasound containing six complete cycles will result in worse axial resolution than one containing three cycles for the same operating frequency. This is illustrated in Fig. 9-3. In practice, most ultrasound pulse echo systems emit beams of 3 to 5 cycles in duration. Normally the medical sonographer cannot alter this condition.

Ultrasound frequency

As ultrasound frequency increases, the wavelength decreases, and if the number of pulses remains constant, the spatial pulse length is reduced. Consequently, at higher frequencies better axial resolution is obtained as shown in Fig. 9-4. Table 9-1 presents typical values of the best axial resolution obtainable as a function of frequency.

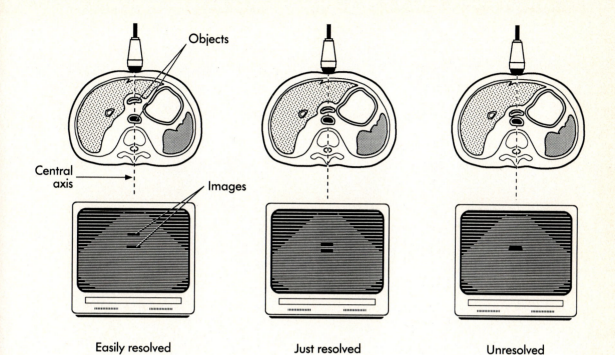

Fig. 9-2 Axial resolution is the ability to resolve closely separated objects or interfaces on the axis of the ultrasound beam.

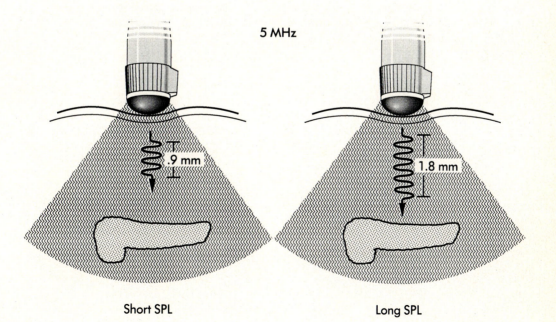

Fig. 9-3 At an operating frequency of 5 MHz, a three cycle pulse will have shorter spatial pulse length and therefore better axial resolution than a six cycle pulse.

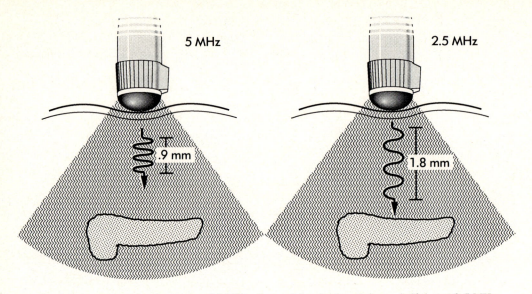

Fig. 9-4 At an operating frequency of 5 MHz, the spatial pulse length is one half that at 2.5 MHz for a three cycle pulse. Therefore axial resolution is better at 5 MHz.

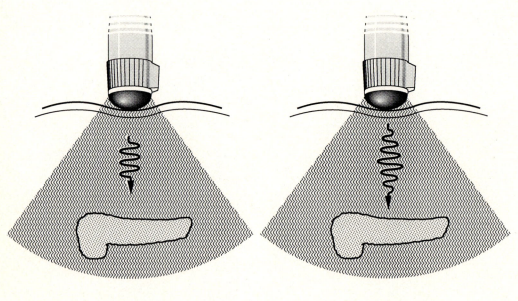

Damped pulse

Undamped pulse

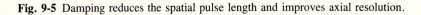

Fig. 9-5 Damping reduces the spatial pulse length and improves axial resolution.

The downside on increasing the operating frequency is the accompanying reduction in depth penetration. For optimum axial resolution one should apply the highest frequency that will also provide the necessary tissue penetration. Furthermore, at high frequency the near field is deeper and the far field more collimated. This contributes to improved axial and lateral resolution in the near field.

Damping factor

To damp an ultrasound beam means to stop the pulse very quickly. Not only is the emitted pulse made sharper by damping, but returning echoes are also sharper. This situation, as illustrated in Fig. 9-5, shows how axial resolution is improved by damping the ultrasound pulse.

Computing axial resolution

The best axial resolution possible with an ultrasound imaging system is termed the **limiting axial resolution.** In the final analysis the axial resolution is determined by the number of cycles in the pulse, damping, and the operating frequency. The number of cycles in the pulse are a consequence of the design of the transducer and its associated electronics. For soft tissue, limiting axial resolution (AR) is given by:

$$AR = 1/2 \; SPL \qquad (9\text{-}1)$$

where SPL = the spatial pulse length.

Spatial pulse length is defined as:

$$SPL = \# \text{ cycles in pulse} \times \lambda; \qquad (9\text{-}2)$$

$$\text{also } SPL = \frac{\# \text{ cycles in pulse} \times v}{f} \qquad (9\text{-}3)$$

Therefore axial resolution is also given by:

$$AR = \frac{\# \text{ cycles in pulse} \times \lambda}{2} \qquad (9\text{-}4)$$

$$\text{and } AR = \frac{\# \text{ cycles in pulse} \times v}{2f} \qquad (9\text{-}5)$$

Example: What is the axial resolution of a three-cycle pulse with wavelength 2 mm?

Answer: AR = 3 cycles × 2 mm ÷ 2
AR = 3 mm

Example: What is the axial resolution of a 3.5 MHz transducer having 4 cycles in each pulse?

Answer: AR = 4 × 1540 m s^{-1} ÷ 2 × 3.5 MHz
AR = 880 × 10^{-6} m
AR = 0.88 mm

Table 9-1 The best axial resolution (mm) obtainable in soft tissue for various operating frequencies and cycles per pulse

Frequency (MHz)	Cycles per pulse			
	2	4	6	8
1	1.54	3.08	4.62	6.16
2.5	0.62	1.23	1.86	2.48
5	0.31	0.62	0.92	1.23
7.5	0.21	0.41	0.62	0.82
10	0.15	0.31	0.46	0.62

LATERAL RESOLUTION

Definition: Lateral resolution is the ability to image closely spaced interfaces perpendicular to the axis of the ultrasound beam.

Symbol: LR

Units: Millimeters (mm)

Lateral resolution, sometimes called **asimuthal or transverse resolution,** refers to the resolution of objects lying in a plane perpendicular to the axis of the ultrasound beam. Lateral resolution is dependent on the size of the transducer element and the operating frequency; it is determined principally by the beam width. Lateral resolution is considerably more important to ultrasonic image quality than axial resolution because it is always worse than axial resolution. **Lateral resolution is usually the parameter that limits image quality.**

Transducer size

When one scans laterally across an object with a single-element transducer, echoes will be returned to the transducer during the entire period it is positioned over the object. If the object were only a point, it would appear as a line whose length is equal to the effective width of the ultrasound beam as seen in Fig. 9-6. If one now scans two point objects separated laterally by a distance greater than the beam width, two straight lines will appear, as in Fig. 9-7. As the point objects are moved closer together the image lines will likewise move closer together until they merge into one line. The minimum object separation distance at which the two objects can still be distinguished is the lateral resolution.

When using a linear array, the specification of lateral resolution is also dependent on the slice thick-

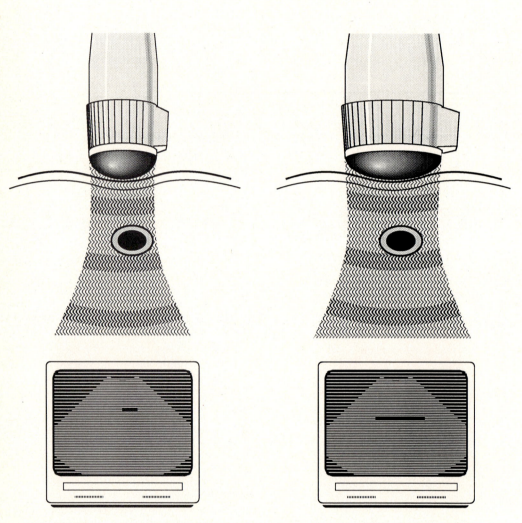

Fig. 9-6 The larger the diameter of the transducer element the worse will be the lateral resolution because the ultrasound beam is wider in the near field.

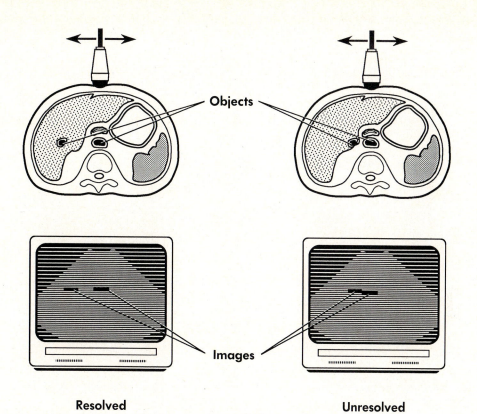

Fig. 9-7 Lateral resolution measures the ability of the ultrasound system to detect closely separated objects lying in a plane perpendicular to the beam axis.

ness of the ultrasound beam as shown in Fig. 9-8. Slice thickness is determined by the design of the transducer array, but is always thinnest at the focal length. Therefore, at the focal length lateral resolution and image contrast are best.

At distances other than the focal length the two-dimensional image represents a greater thickness of tissue, resulting in loss of spatial resolution and contrast.

Beam width

Lateral resolution is approximately equal to the effective beam width, and consequently it depends not only on transducer element size but also on the focusing ability of the transducer. The larger the size of the transducer element the worse the lateral resolution, but for linear arrays that can be electronically focused this is less important. The lateral resolution of a focused transducer array becomes more dependent on the engineering design of the probe. In such cases lateral resolution is defined only at the focus of the transducer. Lateral resolution will be much worse outside of the focal plane. Fig. 9-8 shows how beam width alters lateral resolution for objects on and off the focal plane.

Frequency

Frequency affects lateral resolution in a secondary manner. Recall from Equations 8-7 and 8-8 that higher operating frequency results in a longer near field and a less divergent far field. High frequency therefore produces a narrow beam deeper into tissue with less divergence of the beam as shown in Fig. 9-9. Consequently, the higher the frequency the better the lateral resolution for a greater range in tissue, although the impact is not as great as it is on axial resolution. Unfortunately, high frequency ultrasound does not penetrate tissue as well as low frequency ultrasound.

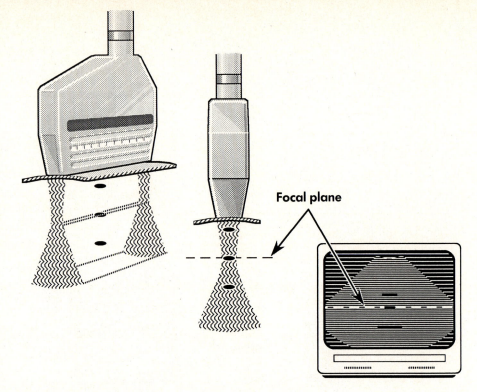

Fig. 9-8 With a transducer array, best lateral resolution and contrast occurs at the focal length because of the thinness of the slice.

Computing lateral resolution

Computing the lateral resolution of an ultrasound image is easy. Lateral resolution is equal to the beam width. If the separation between objects in the same plane is greater than the beam width, they will be resolved. Because the beam width varies with distance from the transducer so will the lateral resolution.

Usually axial resolution is better than lateral resolution, although the two may be comparable in the focal plane. Actual clinical resolution is never quite as good as limiting resolution because of design constraints on the transducer and electronics. In practice, **1 mm axial resolution and 2 mm lateral resolution would be considered good.**

Review Questions: Chapter 9

1. Define or otherwise identify:
   a. Spatial resolution
   b. Contrast resolution
   c. Axial resolution
   d. Lateral resolution
   e. Azimuthal resolution

2. What factor(s) is (are) principally responsible for reducing the contrast resolution of a system?

3. How does axial resolution vary with spatial pulse length, ultrasound frequency, and damping factor?

4. A 5 MHz transducer emits pulses having 5 cycles. What is the axial resolution?

5. Discuss the factors that limit detail on an ultrasound image.

6. Why does lateral resolution vary with depth?

7. For the following, answer true or false.
   a. _____ High frequency (7.5 MHz) transducers are generally focused at a shorter distance than lower frequency (3 MHz) transducers.

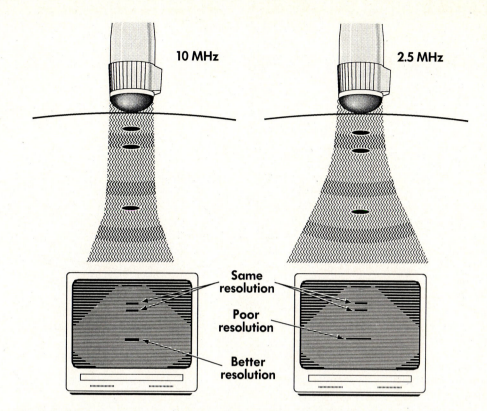

Fig. 9-9 Higher frequency of operation results in a moderately longer near field and less divergence in the far field. Therefore better lateral resolution will result at higher frequency throughout a deeper range in tissue. Of course, high frequencies do not penetrate as far in tissues because of increased absorption.

b. _____ Modern transducers operate at several resonant frequencies by changing the applied voltage on the transducer.

c. _____ Lateral resolution is determined principally by the diameter of transducer.

d. _____ Lateral resolution is better in the Fresnel zone than in the Fraunhofer zone.

e. _____ Increasing the pulse length will increase axial resolution.

f. _____ Higher frequency transducers have better depth penetration than lower frequency transducers.

8. What kind of transducer would be best to image superficial structures that are not behind lung or bone?
   a. Small diameter, high frequency
   b. Small diameter, low frequency
   c. Large diameter, high frequency
   d. Large diameter, low frequency

9. If the spatial pulse length is 10 mm, what is the axial resolution?

10. Lateral resolution can be improved by
    a. increasing the gain.
    b. decreasing the gain.
    c. focusing.
    d. smoothing.

10 Pulse-Echo Imaging

Earlier it was pointed out that there are two types of ultrasound beams, **continuous wave** and **pulsed. Continuous wave** is used for Doppler flow measurements and therapy. **Pulsed** ultrasound is used for imaging, called pulse-echo imaging, and for some Doppler flow applications. Nearly all present ultrasound imaging is done in real-time, with multielement phased transducer arrays becoming the state of the art. Our understanding of such imaging is assisted by knowledge of operational modes that were developed early in the history of diagnostic ultrasound.

Previously there were a number of operational modes available to the sonographer. Two are static imaging modes, A-mode and B-mode. These modes are to ultrasound imaging as radiography is to x-ray imaging. Two are dynamic imaging modes, M-mode and real-time B-mode. Real-time B-mode is normally identified as simply real-time, but real-time is based on B-mode imaging. **Real-time is the fluoroscopy of diagnostic ultrasound.** Each of these modes has found application in diagnostic ultrasound at one time or another, and each has had its own area for special application.

A-mode, which is short for **amplitude mode,** can be used to measure midline shifts of the brain. Most ultrasound units still have the ability for A-mode analysis, but it is only used as a back up to aid in setting up imaging parameters. B-mode or **brightness mode** is the most widely used application and is employed for imaging all areas of the body. Static B-mode has been replaced by real-time B-mode ultrasonography and allows for observation of structures dynamically. M-mode or **motion mode** finds principle application in dynamic evaluation of internal structures.

RANGE EQUATION

The ultrasound transducer generates short bursts or pulses of ultrasound that travel into the patient where they are reflected by interfaces back to the transducer. To properly display the echoes, the instrument uses the time delay between pulse transmission and reception to determine the distance to each interface. This concept, illustrated in Fig. 10-1, can be represented mathematically by a simple expression called the **range equation:**

$$d = vt/2 \qquad (10\text{-}1)$$

where d is the distance to a reflector (m)
     v is the velocity of ultrasound in tissue
     (1540 m s^{-1})
     t is the roundtrip time of the pulse(s)

The 2 in the denominator of the range equation is required to compute only the one-way distance to the interface. It takes only one half of the roundtrip time to reach the reflector.

Example: What is the depth of the interface if the roundtrip time for an ultrasound pulse is 130 μs?

Answer: $d = vt/2$
      $d = 1540 \text{ m s}^{-1} (130 \times 10^{-6}\text{s})/2$
      $d = 0.1 \text{ m} = 10 \text{ cm}$ (the roundtrip distance is 20 cm).

Ultrasound instrumentation uses a velocity of 1540 m s⁻¹ to compute interface depths. Can you show that for a depth in tissue of one cm, a roundtrip transit time of 13 μs is required?

The application of the range equation to each of the pulse-echo techniques is simple and is shown schematically in Fig. 10-2. Here the emission of a single ultrasound pulse from a transducer is reflected from two interfaces each at a different depth in tissue. Notice that three blips occur on the screen of the video monitor instead of two. The first blip is most intense and occurs because of reflection at the transducer-patient interface. The second blip represents the reflection from the first tissue-tissue interface, and it is less intense because of attenuation of the ultrasound beam. Each succeeding tissue interface results in a reflection of reduced intensity

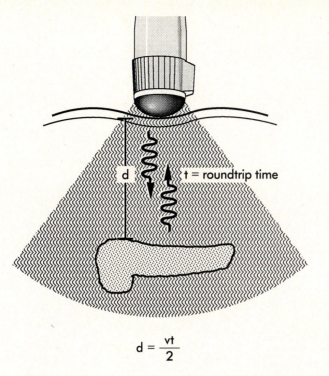

d = roundtrip time

t = roundtrip time

$$d = \frac{vt}{2}$$

Fig. 10-1 The parameters involved in the range equation.

until attenuation is so great that there is no returning ultrasound.

The time required for the pulse to be reflected and returned to the transducer depends on the indicated position of the interface by a blip on the video monitor. Therefore, the ultrasound imager measures three things of importance regardless of the mode of operation: **intensity of reflected echoes, time required for reflection** (which is converted into distance), and the **direction from which the echo was reflected.**

AMPLITUDE MODE

Some of the earliest work with diagnostic ultrasound employed A-mode display. This type of display requires pulse-echo technique. The returning echoes appear on a CRT as a series of blips. The distance between each of the blips is proportional to the distance between interfaces. The height of each blip is proportional to the intensity of each respective reflected echo. Therefore distal reflections produce smaller blips than proximal reflections. These distance and intensity relationships are shown in Fig. 10-2.

The basic components for this type of pulse-echo ultrasound are shown in Fig. 10-3. A-mode ultrasound employs either one or two transducers. In the one transducer mode, shown in Fig. 10-4, the same transducer is used to transmit and receive echoes. This results in a variation in blip size. In a two-transducer application one transducer is used to transmit and the other to receive as shown in Fig. 10-5. For observations in the brain, the transmitting transducer is placed on one side of the head and the receiving transducer on the other side. By alternating the transmit and receive mode of each transducer, blip size remains relatively constant.

The main purpose for employing A-mode is to measure accurately the depth of interfaces and their separation. This type of operation relies on axial resolution which was shown to depend on the spatial pulse length and the frequency of operation.

A-mode instruments are relatively inexpensive, rugged, and easy to use. Except for ophthalmologic applications, they are rarely used anymore. Since depth and separation are the critical measurements, A-mode instruments require frequent, if not daily, calibration. A 2.5 cm thick lucite or polystyrene

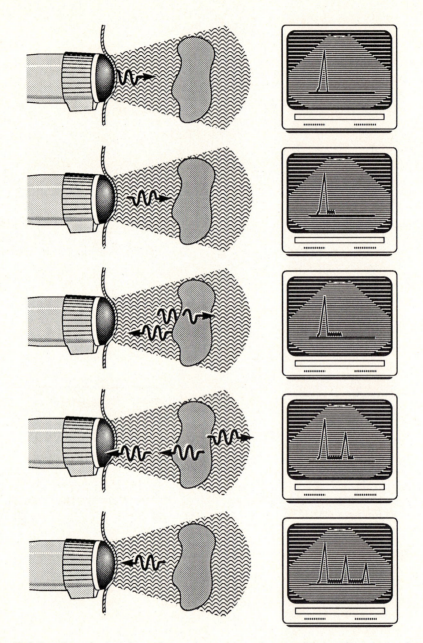

Fig. 10-2 In pulse echo imaging a short pulse traverses tissue until it is reflected at an interface. The time required for the reflected pulse to return to the transducer determines the position of the blip on the CRT face.

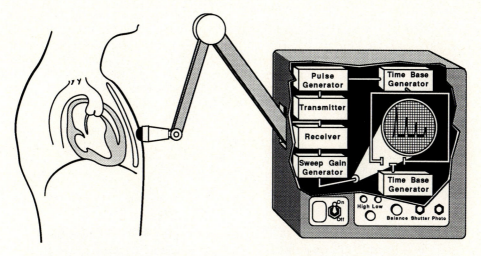

Fig. 10-3 Basic components of A-mode ultrasound and the image display format.

Brain midline

Fig. 10-4 Configuration for one transducer A-mode scanning a brain where no shift of the midline occurs.

block is approximately equivalent to 2 cm of tissue and can be used as a quick calibration tool on a regular basis.

BRIGHTNESS MODE

In the A-mode display the height of the blip is proportional to the intensity of the reflected echo. If that blip is now squeezed down to a dot on the CRT display, its brightness will be proportional to the intensity of the reflected echo. This relationship is shown schematically in Fig. 10-6. This is B-mode or **brightness** mode.

B-mode by itself has little application in diagnostic ultrasound. However, if the spatial position and direction of the ultrasound beam are coupled to the CRT display and the B-mode pulses individually stored while the transducer is moved on the skin, an image will appear that is the summation of many individual B-mode signals. Such an imaging system is shown in Fig. 10-7 and is called **compound B-mode.** Although this type of imaging is correctly identified as compound B-mode, use of the term B-scan is universal. The basic components of a B-scanner are shown in Fig. 10-8.

Fig. 10-5 Configuration for two transducer A-mode scanning showing a midline shift of the brain.

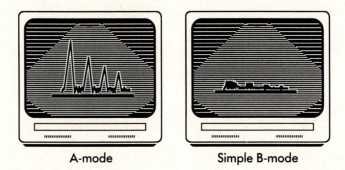

Fig. 10-6 Relationship between A-mode display and simple B-mode display.

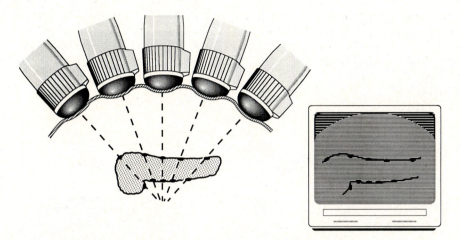

Fig. 10-7 The compound B-scan image is made up of multiple B-scan signals properly oriented.

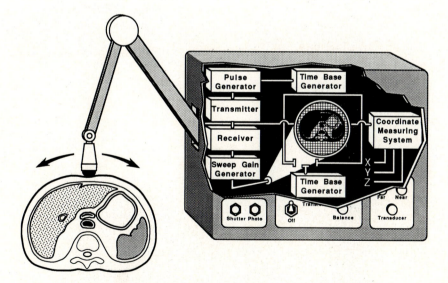

Fig. 10-8 Basic components of a B-scanner and the image format.

Critical to the B-scan image is the proper transducer alignment and indication for position and direction. This alignment is accomplished with either **electric potentiometers** or **optical encoders.** Optical encoders measure changes in visible light with position. The electrical resistance of the potentiometer changes with its position and allows for easy indication of transducer position. Linear and sine-cosine potentiometers are employed. Most B-scan-

ners have a transducer attached to a swinging crane-type arm as shown in Fig. 10-9. This type of apparatus will employ sine-cosine potentiometers. Alternately the transducer can be fixed on a rigid frame as in Fig. 10-10 so that linear potentiometers are required.

Real-time ultrasound does not require the alignment and position indication of the static B-scanner. Consequently, operator skill is less critical with real-

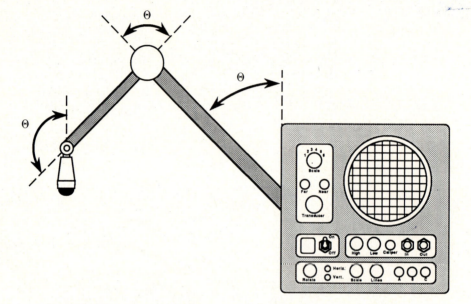

Fig. 10-9 Sine-cosine potentiometers or optical encoders measure angles.

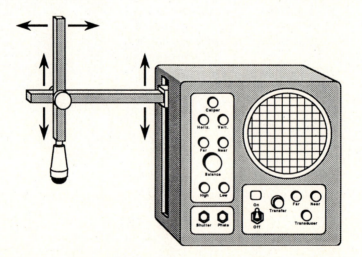

Fig. 10-10 Linear potentiometers or optical encoders measure distance.

time compared to B-scanning. B-scan images require considerable attention and skill on the part of the sonographer because the field of view is usually large, and the image may require several seconds of scan time. Suspension of patient respiration may be necessary.

The effective use of the **windows** of the patient is a more critically required skill of the B-scan ultrasonographer. A window is the best place on the patient-entrance surface which allows an unobstructed path for ultrasound. For example, ultrasound transit through the liver to image the kidney makes use of an unobstructed window. On the other hand, superficial bowel gas represents a poor window because ultrasound will not penetrate to the tissue of interest.

Unlike B-scan images, real-time images are formed automatically many times each second so that image quality is not so dependent on operator skill; a small window is much easier to access. However, since the real-time scanner produces a smaller field of view, more training in anatomy may be required of the sonographer. Real-time scanning also requires a greater understanding of the presentation of that anatomy in relation to the transducer. The tissues closest to the probe are displayed to the top of the image. These may be anterior or lateral tissues even though the patient remains supine.

The ultrasonographer has multiple options for the manner in which the B-mode transducer is manipulated. The transducer can be moved linearly over the patient to provide a rectangular field of view or **linear scan.** Alternately it can be angulated to provide a **sector field** of view. Further still, a combination of both can be employed to provide an **arc scan** or a **compound scan.** These transducer motions are illustrated in Fig. 10-11. The resolution of a B-mode image is determined not only by the transducer characteristics described in Chapter 7, but also by the mechanical-electrical linkage that determines the position and direction of the ultrasound beam.

Early B scanners incorporated a bi-stable or black

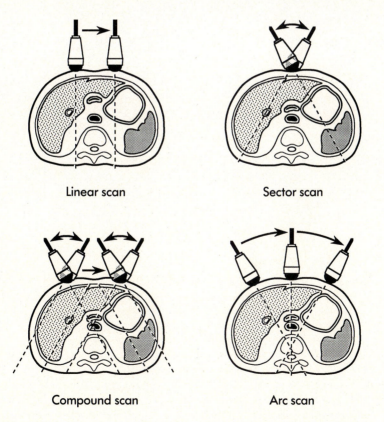

Linear scan

Sector scan

Compound scan

Arc scan

Fig. 10-11 Types of transducer motions used in B-mode ultrasound imaging.

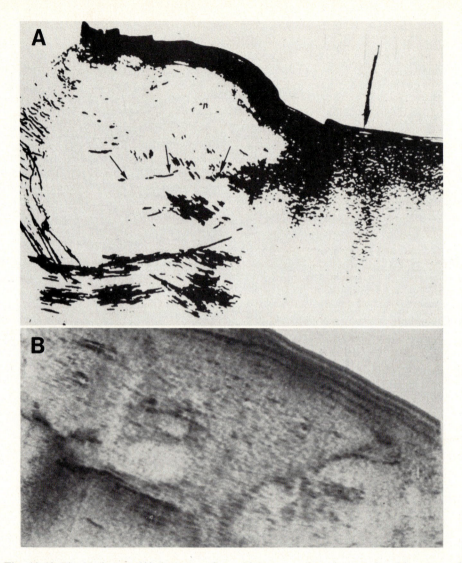

Fig. 10-12 Bistable images **(A)** that were of very high contrast have been replaced by gray scale images **(B).**

and white display mode. This type of display showed very high contrast, only bright spots and lines on a blank screen with no intermediate intensities on the face of the CRT. Gray scale display mode was subsequently developed, and it provides a range of intermediate display intensities. Figure 10-12 demonstrates the difference between (A) bi-stable and (B) gray scale ultrasound images.

Gray scale is universally employed now, and with the integration of computers the contrast scale has been considerably broadened. Gray scale is now pos-

sible by using a small dedicated computer and a special type of CRT tube called a **scan converter.**

Analog and digital scan converters have been used in diagnostic ultrasound imaging, but the use of digital scan converters predominates. Currently, all diagnostic ultrasound systems incorporate digital scan converters and microprocessors. With a digital scan converter, data are stored in digital form and displayed as an image matrix similar to computed tomography and magnetic resonance imaging. This allows digital ultrasound images to be postprocessed

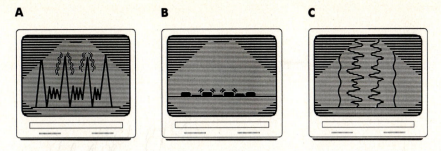

Fig. 10-13 If the A-mode contains moving interfaces and is converted to a B-mode display so that the vertical axis is time driven, M-mode display results. **A,** A-mode. **B,** B-mode. **C,** M-mode.

for more flexible image analysis and interpretation. Unfortunately, when such images are photographed the gray scale is less than that of the human eye. Scan converters and image processing will be discussed in Chapter 12.

MOTION MODE

If a pulse-echo transducer is positioned over the heart and operated in A-mode, the image will contain a number of blips, some stationary and some vibrating back and forth as in Fig. 10-13, *A*. The vibrating blips represent interfaces that are in motion relative to the transducer. The magnitude of the vibration of each blip represents the degree of movement of the tissue interface.

If this A-mode display is now converted to a single B-mode line, the image would contain a series of dots, some fixed and some moving as in Fig. 10-13, *B*. The fixed dots are related to the stationary interfaces and the moving dots are related to the moving tissue interfaces. If the vertical axis of the display is now driven as though it were a strip of paper in a strip chart recorder, a tracing of the dots will result as shown in Fig. 10-13, *C*. The stationary dots trace out a straight line while the moving dots trace a regular pattern according to the motion of the associated tissue interface.

This type of ultrasound display is commonly called M-mode, but it is also sometimes referred to as TM-mode (time-motion mode), PM-mode (position-motion mode), and UCH (ultrasonic cardiography) since its principle application is to monitor the heart.

On the M-mode display, the horizontal axis records the depth into tissue and the vertical axis measures the time interval between successive heartbeats. An advanced feature of M-mode display allows it to be synchronized with an electrocardiogram tracing for even better evaluation of cardiac function.

Review Questions: Chapter 10

1. Define or otherwise identify:
   a. A-mode
   b. The range equation
   c. B-mode versus compound B-mode
   d. Motion mode

2. Ninety-four μs are required for an ultrasound pulse to be emitted, travel through soft tissue, and the echo received. What is the depth of the interface?

3. Describe the process of acquiring a static B-mode image. What were the major disadvantages of static B-mode imaging?

4. Describe the characteristics and use of an M-mode display.

5. Describe the process of obtaining an A-mode image, and explain what information is presented in the display.

6. The round trip travel time for an ultrasound pulse in soft tissue is 30 μs.
   a. How deep did the pulse travel?
   b. What is the total distance traveled?

7. How long does it take for a pulse to travel a round trip distance of 18 cm in soft tissue?

11 Real-Time Transducers

The principle drawback to the static B-scanner is the time required to construct an image. The transducer must be manually manipulated (scanned) over the patient's skin by (1) angling to produce a sector scan, (2) translating to produce a rectangular image, or (3) using a combination of both, all of which take considerable time. An improvement over B-scanning is real-time imaging.

The term real-time is borrowed from computer science and implies instant computation or, as in diagnostic ultrasound, instant imaging. Real-time imaging became the scanning mode of choice with the introduction of the **digital scan converter** and the **microprocessor.** The principle advantage of real-time is the ability to obtain images instantly and continuously of both static and moving internal structures. While moving the probe over the patient, one is able to view essentially all planes and all anatomy, assuming there is no interference from gas or bone.

These devices have allowed almost instantaneous reception, processing, and display of pulse-echo information. Semiconductor memory has become so large and inexpensive that multiple microprocessors are now used in most real-time imagers, allowing wide control of focusing, steering, and amplification of the ultrasound beam. This computer technology also allows for increased information density, providing images of superior quality to those obtained with static B-scanning.

Real-time ultrasonography is to static B-scanning as fluoroscopy is to radiography in x-ray imaging. The box on page 89 summarizes the types of real-time imagers available. There are several distinct advantages to real-time imaging when compared to static B-scanning.

1. Image acquisition is quicker.
2. It is easier for the sonographer to produce an image.
3. The image continuously updates with probe position.
4. The movement of internal structures is imaged.
5. Portable imaging is possible.

The ability to observe internal motion is important in several instances. Observing fetal motion indicates a live fetus. Color flow imaging can be helpful in distinguishing arteries, veins, and ducts. Peristalsis (bowel motion) provides an important distinction between fluid-filled bowel and extraluminal fluid.

If a static image is required for study, it is easily obtained as a freeze frame using a memory device. The static images can be later transferred to hardcopy.

Newer systems digitize the real-time image which allows for postprocessing. Table 11-1 compares some of the characteristics of a real-time imager to those of a static B-scanner. The real-time image has a number of instrument-related characteristics that distinguish it from a static B-scan image.

Characteristics of real-time images

Whether mechanical or electronic, real-time images are produced with discrete lines in the image. The number of lines in an image and the rate at which they are produced are principle characteristics of the image. The maximum depth of imaging places limits on these characteristics.

Line density (LD). Line density is the number of lines in an image. Fig. 11-1 shows that these lines can be parallel in the image of a linear array scanner, or they can radiate from a point in a sector scanner.

In a mechanical sector scanner, the LD is determined principally by the design of the imager. The most important characteristic is the number of ultrasound pulses emitted during each sweep of the transducer. Fig. 11-2 illustrates how image quality is improved with higher line density. In a mechanical

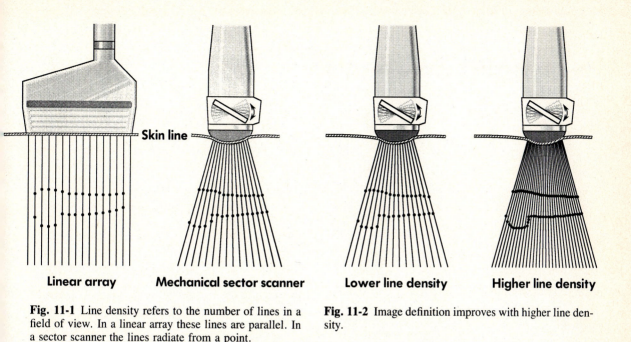

Fig. 11-1 Line density refers to the number of lines in a field of view. In a linear array these lines are parallel. In a sector scanner the lines radiate from a point.

Fig. 11-2 Image definition improves with higher line density.

Table 11-1 Characteristics of real-time imagers and B-scanners

Real-time	Static B-scan
Continuous image production	Several seconds per image
Small field of view	Large field of view
Automatic image formation	Manual image formation
Post processing of image	Fixed image
Image quality determined by scanner	Image quality determined by operator and scanner
All planes imaged easily	Restriction on imaging planes by articulated arm
Some equipment compact and portable	Large equipment size requiring cart
Organ motion can be viewed	Organ motion degrades image

TYPES OF REAL-TIME IMAGERS

Mechanical
Short fluid path
Long fluid path
Rotating
Oscillating

Electronic arrays
Linear array (rectangular scan)
Phased array (sector scan)

Annular array

sector scanner, the LD is determined by the number of transducer elements and the pulsing sequence of each. Ultrasound probes with transducer elements tightly packed or pulsed more frequently will produce a higher line density and better image quality as shown in Fig. 11-3.

The rectangular image usually covers 5 to 10 cm. The sector image will typically appear as a 45° or 90° pie shape.

In general image quality improves with increasing scan-line density. However, such improvement does not come without cost in design and fabrication, especially for phased array sector scanners. The lin-

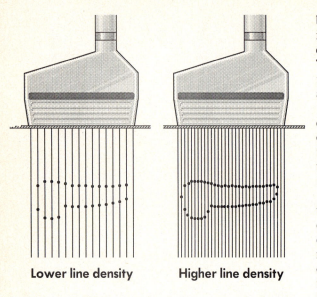

Lower line density   **Higher line density**

Fig. 11-3 Image quality improves as the number of transducer elements is increased.

ear array may produce blank lines between the lines of the image. This striped appearance can be reduced by computer-assisted averaging between the lines to smooth the image. With a sector scanner the process is more difficult. The degree of interpolation must continuously increase with depth to compensate for the diverging lines of the image.

Frame rate (FR). Each real-time image is made of a series of frames. The frequency with which frames are presented is called the frame rate (FR), which is often operator selectable. Frame rate is particularly important for real-time imagers when used to evaluate moving structures. The integration time of the eye is approximately 200 ms. Therefore, one would expect that a FR less than about five frames per second (5 f s^{-1}) would result in flicker. It does! At 5 f s^{-1} the flicker is very distinct. Flicker does not disappear until the FR exceeds approximately 20 f s^{-1}.

High frame rates are better for motion studies such as cardiac imaging, but image quality is compromised by less focusing and lower line density. Frame rates of 30 f s^{-1} are common, but lower FRs may be necessary to accommodate the LD and the maximum depth of the image.

Frame rates as high as 60 f s^{-1} can be employed.

but higher FRs are limited by the velocity of sound in soft tissue. The FR cannot be so fast that returning echoes are detected after the next pulse is emitted. Therefore, the upper limit on the FR is also determined by the LD and the pulse repetition frequency, as well as the number of focusing planes.

Pulse repetition frequency (PRF). The pulse repetition frequency, sometimes termed the **pulse repetition rate,** was introduced in Chapter 8 as the frequency of emission of ultrasound pulses. Recall that PRF refers to the number of pulses emitted each second by a transducer element.

Following emission of an ultrasound pulse, the transducer element behaves as a receiver of pulse echoes. It will not emit a successive pulse until all echoes have been received. The time required to receive all echoes depends on the maximum tissue depth to be imaged and the number of lines in the image. The deeper one wishes to image, the longer it will take to receive echoes, and therefore the PRF will be lower.

The PRF identifies the rate at which transducer elements are energized; therefore the PRF is the product of the frame rate and the number of lines per frame. Multiplane focusing also takes time. The number of lines per frame, the line number, is equal to the number of transducer elements. Therefore,

PRF (Hz) = frame rate × line number   **(11-1)**

Example: A 64-element transducer probe is operated at 30 fps. What is the pulse repetition frequency?

Answer: PRF = f s^{-1} × 64 lines = 1920 Hz

All real-time imagers have inherent limitation on image quality determined by the frame rate and pulse repetition frequency. Normally the only constant is the line number (LN) determined by the number of transducer elements. As manufacturers attempt to improve image quality by using more transducer elements, the frame rate must be reduced because of the limiting pulse repetition frequency.

Example: It is determined that the maximum PRF for a given probe is 3.6 KHz. To improve lateral resolution the manufacturer packs 256 transducer elements into the probe. What is the maximum possible frame rate?

Answer: FR = PRF ÷ 3600 ÷ 256 = 14.1 f s^{-1}

This design will result in flicker, which may be objectionable for motion studies.

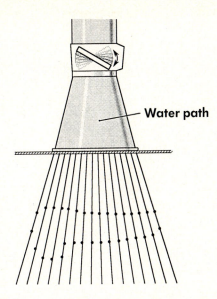

Fig. 11-4 Some mechanical scanners incorporate a water path between the transducer and the patient's skin. The increased depth of imaging results in a lower pulse repetition frequency.

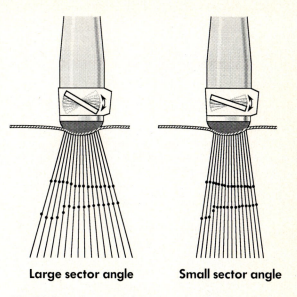

Large sector angle   **Small sector angle**

Fig. 11-5 As the angle of the sector is reduced the image quality improves but the field of view is restricted.

Some mechanical scanners incorporate one or more transducers separated several centimeters from the skin by a water path. This feature is diagrammed comparatively in Fig. 11-4. The result of the added path length is a lower PRF for these scanners.

Maximum depth of image. A transducer probe designed for imaging deep structures will have a lower PRF than one designed for shallow structures. A properly designed imager will not allow production of one line of the image until the preceding line is completed. The preceding line cannot be completed until the pulse of ultrasound has been reflected from the deepest structure and the echo returned to the transducer. Therefore, **the PRF is limited by the maximum depth of the image.**

Sector angle. When evaluating the image quality of a sector scanner, the angle of the sector is a primary characteristic. For a fixed LD, the image will appear sharper as the sector angle is reduced (Fig. 11-5). The disadvantage of a small sector angle is the reduced field of view.

Interdependence of these characteristics. Each of these imaging characteristics is dependent on one or more of the others. Optimum imaging requires a proper balance of each of these characteristics. The box to the right provides a summary of these relationships.

RELATIONSHIPS AMONG IMAGING CHARACTERISTICS*

If the line density is increased,
1. image quality will improve.
2. the pulse repetition frequency will be lowered.
3. the maximum depth of image will be less.
4. the frame rate will be lower.

If the pulse repetition frequency is increased,
1. image quality will improve for objects in motion.
2. the line density may limit the increase of PRF.
3. the maximum depth of image will be less.
4. the frame rate will increase.

If the frame rate is increased,
1. image quality will improve for objects in motion.
2. the line density may limit the increase of FR.
3. the maximum depth of image will be less.
4. the pulse repetition frequency will increase.

If the maximum depth of image is increased,
1. the image will extend over a greater depth.
2. the line density will remain fixed.
3. the pulse repetition frequency will be reduced.
4. the frame rate may be reduced.

If the sector angle is increased,
1. image quality will be degraded.
2. the field of view will increase.
3. the line density, pulse repetition frequency, frame rate, and maximum depth of image will remain unchanged.

* Assumes no compensating adjustment.

These relationships hold for both approaches to real-time imaging, mechanical and electronic. Mechanical real-time transducers appeared first but have been supplemented by electronic real-time transducer arrays.

Mechanical real-time transducers

The development of mechanical real-time transducers has involved several different designs. The principle of each is to cause an ultrasound beam to sweep repeatedly across the field of view. The result is a sector scan image through a 45° to 90° arc that is refreshed with every sweep of the transducer element. Most designs use a single transducer that is oscillated inside a fixed case as shown in Fig. 11-6.

Mechanical imagers are not manufactured to be in contact with skin, although some early models were nearly in contact through a membrane. Normally the transducer element is at a distance from the skin so that the ultrasound beam must pass through a fluid-filled path that restricts the depth of image as was shown in Fig. 11-4. They are identified as short or long fluid paths depending on the distance between the transducer and the skin. A short path transducer would be less than 1 cm while a long path transducer greater than 1 cm.

Short water path imagers suffer from easy production of **reverberation** (see Chapter 14) caused by an acoustic impedance mismatch at the patient's skin. Use of fluids other than water can minimize this effect. Longer path scanners are designed to remove reverberation artifacts by making the length of the fluid path equal to the patient depth to be imaged. Unfortunately, this often results in an unacceptable loss of ultrasound intensity.

An alternate arrangement, though not often used, is shown in Fig. 11-7. In this scheme the transducer element is fixed within the probe, and a reflecting surface is caused to oscillate. Again, the result is a sector scan image that is continuously refreshed.

Several manufacturers have produced a rotating multielement transducer design as seen in Fig. 11-8. This design incorporates three or four transducers rotating on a common shaft and spaced either 120° or 90° apart respectively. The wheel on which the transducers are mounted is 2 to 5 cm in diameter and rotated by an external motor. Microswitches allow only the transducer element aligned with the acoustic window to transmit pulses and receive echoes. The result is again a sector scan, except that the image is refreshed more rapidly.

There are subtle advantages to oscillating or rotating imagers that can influence the coupling fluid length designed into the probe. As a general rule, short fluid length probes are designed with rotating transducers rather than with oscillating transducers. Rotating transducers have a constant velocity as they sweep across the field of view. On the other hand, oscillating transducers come to a complete stop twice

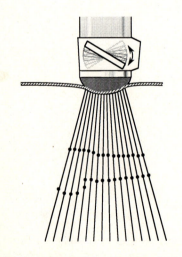

Fig. 11-6 By mechanically oscillating a single transducer inside a fixed probe a real-time sector scan will be produced.

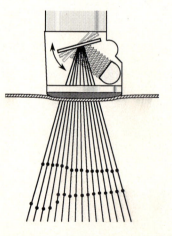

Fig. 11-7 This mechanical real-time transducer incorporates an oscillating reflective surface in the probe.

in each cycle resulting in a changing probe sensitivity across the field of view. Both types of transducers have potential vibration artifacts caused by mechanical motion, but these have been largely accommodated by design.

Mechanical real-time transducers are losing ground to electronic transducer arrays. Still, a distinct advantage to many mechanical designs for use in the abdomen is the hemispherical, concave housing that is often incorporated into the design. This provides excellent acoustical coupling with the skin, especially for imaging between the ribs. However, this design is a disadvantage in neonatal heads because the transducer cannot be brought into complete contact with the skin. Electronic transducer arrays are often longer and flat, requiring a larger area of pliable tissue, such as the abdomen, for good acoustic coupling. Smaller sizes however can now be used for inaccessible areas such as cardiac work or neonatal heads. If the far field is important, as in the deep abdomen, then larger size transducers are used.

Electronic real-time transducer arrays

Multiple element transducers fabricated into a linear array, such as the one shown in Fig. 11-9, produce essentially instantaneous images. The multielement array usually contains from 64 to 256 small individual piezoelectric crystals measuring 1 to 2 mm wide by 5 to 10 mm long positioned in a line. The complete array will measure 4 to 10 cm in length. The number of lines of information in the field of view, the line density (LD), is related to the number of transducer elements. The number of lines is usually a few less than the number of elements. The number of images displayed each second is the frame rate (FR). The frame rate is usually set at 20 f s^{-1} or higher so that flicker is not noticeable.

Linear array. The manner in which these transducer elements are energized differs from design to design. **Sequential activation** is shown in Fig. 11-10. When an individual transducer element emits a pulse, it must remain energized long enough to receive the reflected ultrasound echoes before the next element is energized. In this transducer, element number one is energized first while all others are off. A pulse is emitted and echoes returned to element one; then element number two is energized. Each element behaves independently, emitting ultrasound and receiving these echoes before the next element is energized.

The main problem with sequential activation is the very poor beam pattern that results. In this case, because of the small size of each element, they radiate a nearly spherical beam that is very poor for imaging. This is illustrated in Fig. 11-11, *A*. To overcome this, several transducer elements are activated simultaneously as in Fig. 11-11, *B*. This results in a more collimated beam and flatter wavefront, similar to a single large transducer.

The scheme of simultaneous activation of several transducer elements is called **segmental activation,** and it is shown in Fig. 11-12. In a segmental linear array, four to eight contiguous transducer elements are activated simultaneously. Each pulse of ultrasound from this small group of elements results in

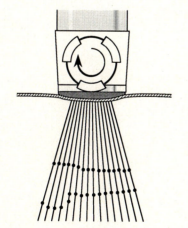

Fig. 11-8 A real-time transducer probe that contains three transducers rotating on a common shaft.

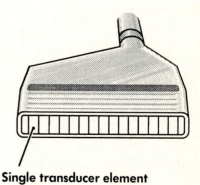

Single transducer element

Fig. 11-9 Real-time electronic probes contain up to 256 individual transducer elements aligned linearly.

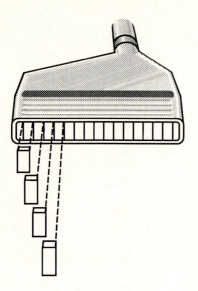

Sequential linear array

Fig. 11-10 Sequential activation of transducer elements occurs one at a time and in order.

A

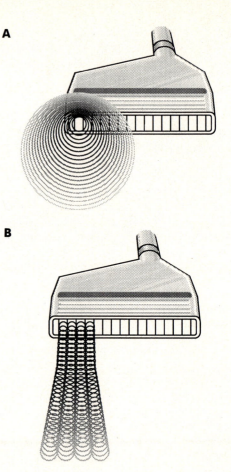

B

Fig. 11-11 A, Sequential activation of transducer elements results in a spherical wavefront that degrades the image. **B,** Segmental activation results in a more collimated beam with a flatter wavefront.

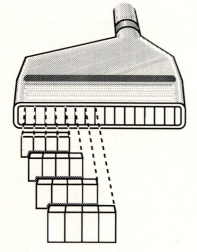

Segmental linear array

Fig. 11-12 Segmental activation of transducer elements involves nearly simultaneous excitation of several elements to form a pulse.

one scan line. In a typical system, transducer elements number one to five will be energized for one pulse. The second ultrasound pulse will be emitted by transducer elements two through six, which again will provide one scan line of information. Next, transducer elements three through seven are energized to produce an additional scan line of information. This sequence is shown in Fig. 11-13. The entire array will be activated each 1/15 s to 1/60 s producing frame rates of 15 to 60 f s^{-1}. The line density is related to the number of elements in the array, the number of elements energized in the pulse, and the pulsing sequence.

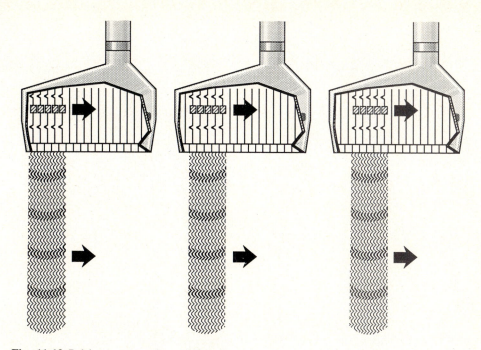

Fig. 11-13 Pulsing sequence for real-time linear array involves segmental activation of multiple transducer elements.

Fig. 11-14 This family of ultrasound probes contains a convex model *(arrow)* and several additional types. (Courtesy Advanced Technology Laboratories).

The segmental linear array uses each transducer element several times in a frame, thereby increasing the signal to noise ratio for that image. This is a distinct advantage over the sequential array and consequently results in a significantly better image with no loss of resolution.

These linear arrays result in a rectangular image format, the lateral dimensions of which are slightly less than the physical width of the transducer array, usually 5 to 10 mm. The length of the transducer array is usually several cm longer than the rectangular scan width. The depth of penetration is a function of the ultrasound frequency and the manner in which each element of the array is energized.

A recent addition to the linear array family is the convex array probe, an example of which is shown in Fig. 11-14. Such probes result in a sector field of view with resolution equal to that of a linear probe. Fig. 11-15 illustrates another advantage to the curved array. Returning echoes are more readily detected when they fall perpendicularly rather than obliquely on each transducer element. Overall, the sensitivity is increased.

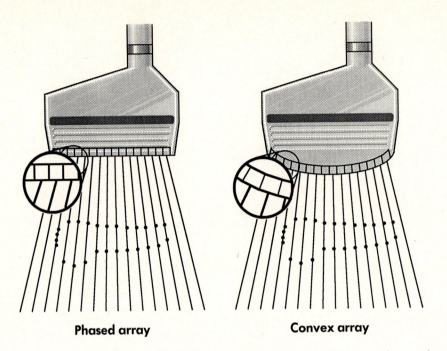

Phased array **Convex array**

Fig. 11-15 The convex array transmits and receives ultrasound at right angles to the transducer surface. This preserves resolution at the sides of the sector scan.

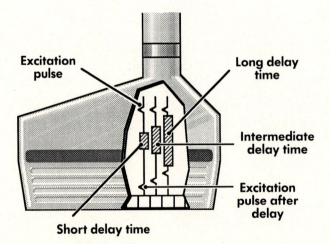

Fig. 11-16 A phased array transducer assembly has electronic delay times for each element that can be varied according to the needs of the examination.

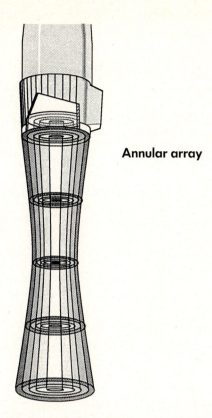

Annular array

Fig. 11-17 An annular transducer array is fabricated with concentric crystal elements, each having the shape of an annulus. The ultrasound beam can be focused and is three-dimensional.

Phased array. In a phased array transducer an electronic delay is incorporated into the excitation of each transducer element. This delay, shown in Fig. 11-16, allows for a precise time of energization of each element to provide for focusing and steering of the ultrasound pulse. Electronic steering is not possible with a linear array. Although there are some proprietary modifications, timing of pulses to each of the phased array transducer elements determine the exact shape and direction of the beam.

The sector scan format is accomplished by electronic circuitry incorporating a delay time to precisely sequence the excitation of an ultrasound pulse and the echo reception by each transducer element. By changing the excitation time from one frame to the next, that is, the **phase** of one transducer element to the next, one can not only form a sector scan but also determine the angle of the sector. The use of

phased array in this fashion is termed **steering** the ultrasound beam. **Focusing** such an ultrasound beam using delay times is also possible.

Phased array imaging produces a sector scan having a maximum sector angle of 90°. The sector is produced by phased excitation of the transducer elements which results in steering the ultrasound wavefront. The line density (LD), frame rate (FR), number of focusing planes, and maximum depth of image are independent and are often operator selectable.

To accommodate the sector scanning nature of the phased array, the total transducer size is smaller than that of a conventional linear array. Phased array probes are generally 1 to 3 cm in length. The individual transducers are so small they begin to radiate isotropically as a point source, and this increases the noise of the system and somewhat degrades the image.

The pulse repetition rate is determined by the size of the transducer elements and the rate at which they are energized. The frame rate is a function of overall transducer size and is equal to the pulse repetition rate. The maximum depth of an image is a function of transducer frequency and electronic delay time circuitry that allows for variable depth of focus.

ANNULAR ARRAY

A way to vary the focus electronically is to fabricate circularly shaped transducer elements as seen in Fig. 11-17. Such a transducer has five to ten concentric piezoelectric crystals that are energized sequentially from inside out, resulting in circular symmetry of the beam. The data obtained is three dimensional even though only two dimensions can be visualized. Energizing the elements in phase provides for variable focusing of the ultrasound beam along the central axis but does not permit beam steering.

Steering the ultrasound beam is still done mechanically. While the transducer elements are energized sequentially, the entire transducer assembly is oscillated from side to side to produce a sector scan.

SPECIALTY ULTRASOUND PROBES

The newest generation of ultrasound transducers includes small transducer devices a little larger than pencils. These are designed for insertion into the rectum and vagina for evaluation of these and adjacent organs. Some designs incorporate a single transducer element that mechanically oscillates through 45° to 90° to produce a sector scan image.

A

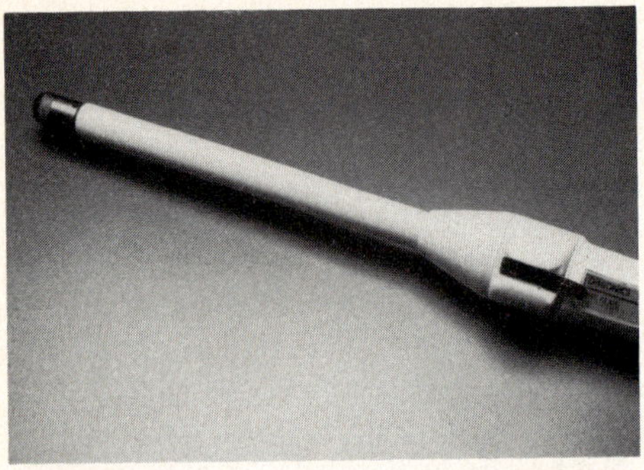

B

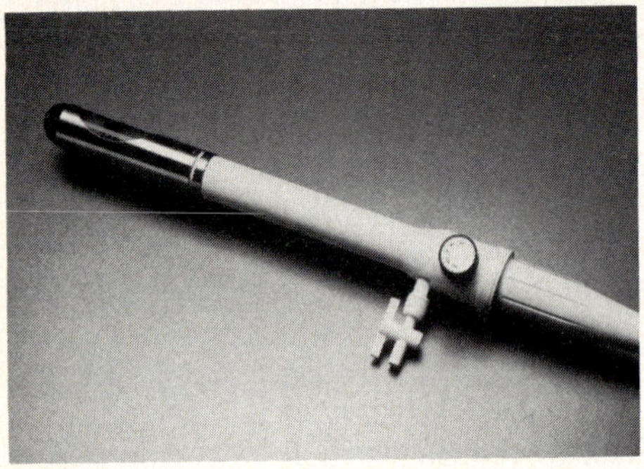

Other designs incorporate convex and linear array or phased array technology.

Examination of the prostate is relatively easy with a probe designed for rectal insertion (Fig. 11-18, *A*). Probes designed for intravaginal use are similar in design but relatively shorter and smaller than the rectal probe (Fig. 18, *B*). A schematic diagram of such intraluminal probes is shown in Fig. 11-19. The features of such probes are small crystals, flexible casing, and a wide field of view.

Table 11-2 presents a brief summary of the three principal types of real-time imagers. This summary is intended as a generalization. Manufacturers are continually improving on each of these designs.

FOCUSING REAL-TIME ULTRASOUND

Focusing of early model and single element transducers is done mechanically. Either the transducer face is shaped or an acoustic lens is employed as discussed previously in Chapter 8. Many linear arrays are also focused mechanically in one dimension with an acoustic lens as seen in Fig. 11-20.

Fig. 11-18 These are high frequency multi-plane probes designed for the following. **A,** This probe is designed for rectal insertion for examination of prostate. **B,** This probe is designed for vaginal insertion for pelvic examination and first trimester obstetrics studies. (Courtesy Diasonics).

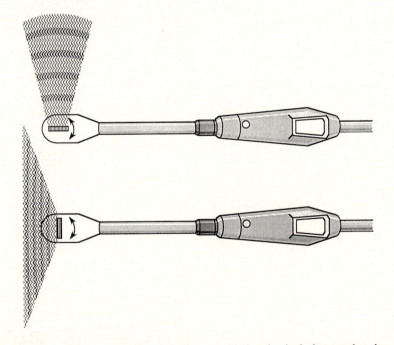

Fig. 11-19 Schematic drawings of representative intraluminal ultrasound probes.

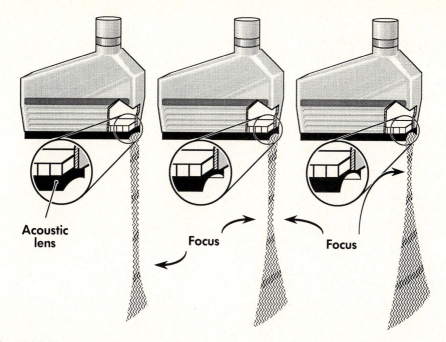

Fig. 11-20 Linear arrays are usually focused perpendicular to the array length with acoustic lenses placed in front of the transducer elements.

Table 11-2 Advantages and disadvantages of the three principal types of real-time imagers

Advantages	Disadvantages
Mechanical	
Less expensive	Potential mechani-
Simple to use	cal failure
Light	Fixed focus
Portable	Cannot perform
	color flow
	Limited frame rate
Phased array	
No moving parts	Expensive
Multipurpose	Relatively large and
Color flow capability	heavy
Multiple focal planes	Potential side lobes
Variable focal depths	degradation
Annular array	
Thinner slice at focal point	Potential mechani-
Variable focus	cal failure
	Limited resolution
	off focal plane
	Color flow difficult
	Limited flow rate

Electronic focusing

Linear array transducers are also focused electronically. They take advantage of delay time electronics made possible by microprocessors. Excitation pulses to individual transducer elements can be precisely controlled by varying delay times. Fig. 11-21 shows how this works.

The use of delay times to focus an ultrasound beam is called **electronic focusing.** The pulsing delay times are measured in nanoseconds (10^{-9}s), while the time a transducer element is active is measured in hundreds of microseconds (10^{-6}s).

Dual focusing used in linear arrays is shown in Fig. 11-22. The complement of an acoustic lens and electronic focusing results in two dimensional focusing. The dimension determined by the acoustic lens is the **slice thickness** in the imaging plane.

Here the outside transducer elements are energized first while excitation of the middle elements is delayed. This effectively causes the lateral lines of the ultrasound beam to be pinched into the middle. The result is to move the near field-far field transition closer to the transducer face. The greater the timing delay difference in excitation of the individual elements the shorter the focal length will be.

When the ultrasound beam is formed from a mul-

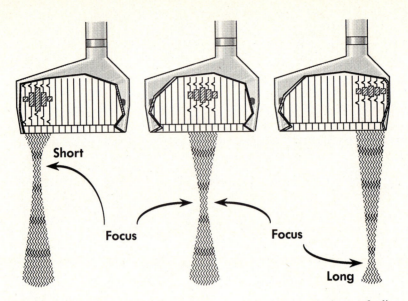

Fig. 11-21 Delaying the time of element excitation results in focusing the beam of a linear array.

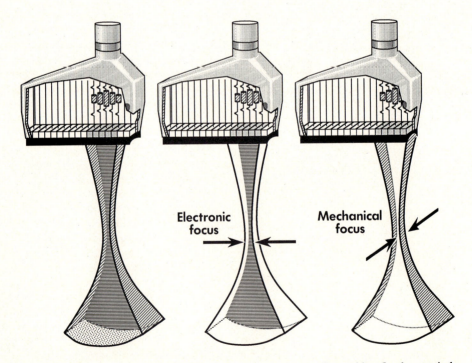

Fig. 11-22 A linear array is focused perpendicular to the image plane with a fixed acoustic lens. Focusing in the image plane is accomplished electronically by delayed pulsing of individual transducer elements.

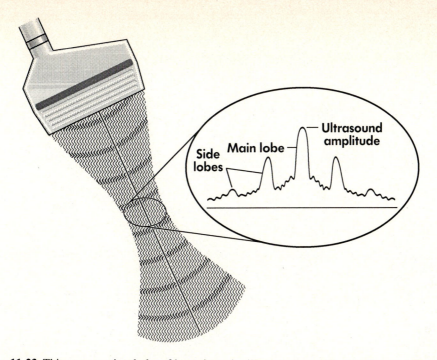

Fig. 11-23 This cross-sectional plot of beam intensity illustrates how objectionable side lobes can degrade lateral resolution.

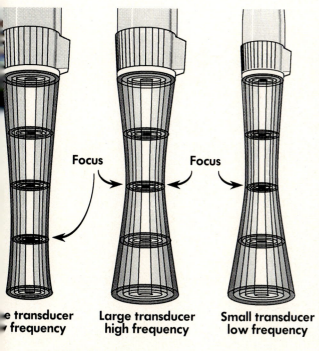

Fig. 11-24 The focal length of an annular ultrasound probe is determined by the size of the annular elements and the operating frequency.

tielement array, objectionable side lobes may be produced. These can reduce lateral resolution and degrade image quality. Fig. 11-23 shows an actual beam profile with its side lobes superimposed on the ideal beam profile. Side lobes are a consequence of vibrations along the length or width of the crystal as opposed to the thickness mode vibrations discussed. The greater the side lobes the more the image contrast and lateral resolution will be degraded.

The loss of image contrast due to side-lobe interference can be improved by **apodization.** Apodization is a term used in optical physics to describe the scattering of a collimated light beam by diffraction of the side portion. Engineering techniques are designed into linear arrays so that the sensitivity of the peripheral transducer elements is reduced. These elements are most responsive to side-lobe interference.

Annular arrays can be similarly focused by electronic delay of excitation pulses. Fig. 11-24 shows the beam profile from a simple annular array. The focal length of this probe is determined by the size of transducer elements and the operating frequency. By incorporating excitation pulse delay electronics, as shown in Fig. 11-25, the focus of the annular ultrasound beam can be improved.

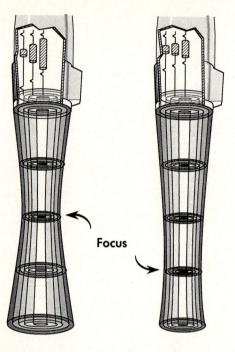

Fig. 11-25 The focal length of an annular array can be changed by incorporating excitation pulse delay electronics.

Dynamic electronic focusing

Applications of dynamic excitation pulse delay electronics results in **dynamic focusing.** In this technique the focal length of the transducer array is changed from one frame to the next. This effectively extends the imaging zone over a greater range of tissue. Although image quality is improved by dynamic focusing, some reduction in frame rate or line number may be necessary. Consequently, some reduction in contrast resolution can also be expected. Most manufacturers now provide phased array ultrasound probes with dynamically variable focal lengths. The principal result, of course, is improved lateral resolution at various depths in tissue.

STEERING THE ULTRASOUND BEAM

The linear transducer array produces a rectangular field of view that is focused mechanically and electronically. The phased array transducer contains fewer smaller transducer elements and produces a beam that can be steered and focused. The phased array produces a sector scan because essentially all

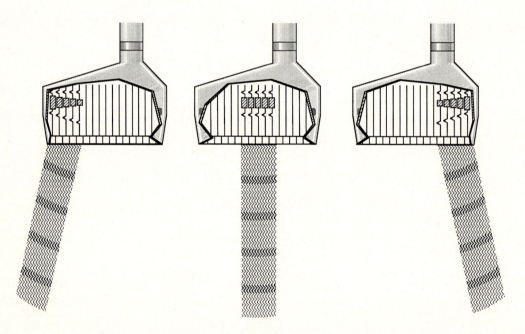

Fig. 11-26 Dynamic excitation pulse delay electronics can be used to steer the ultrasound beam.

of the transducer elements are activated at the same time producing one line of the image. The line density of a phased array image is therefore equal to the number of times the array is pulsed per frame. By slightly altering the delay times and phase of each element, the beam will be steered as seen in Fig. 11-26.

Steering the ultrasound beam is another characteristic made possible by the microprocesssor. Much like focusing, steering the ultrasound beam is accomplished by precisely timing the energizing of each transducer element. In much the same way that electronic delay times produce a focused beam, the delay times can also be used to steer the beam. Fig. 11-27 shows how the angle of the sector scan is controlled. The greater the phase difference between excitation of the transducer elements, the larger the sector angle will be.

In addition to steering the ultrasound beam to produce a sector scan, the phased array can be focused by a very complicated series of delay times to each transducer element (Fig. 11-28). As with the linear array, differences in delay times are measured in nanoseconds while each line generating pulse is spaced by milliseconds. Unlike the linear array, the phased array involves the constant changing of the phase among elements since all elements participate in forming each line of the image.

Phased array transducers can produce dynamically focused ultrasound beams. This feature is usually operator selectable, and there can be a limitless choice of depths of focus. However, as more focal planes are selected the frame rate will be reduced.

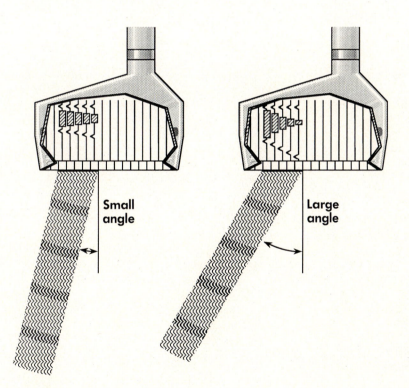

Small angle

Large angle

Fig. 11-27 Precisely timing the electronic delay times to a phased array can produce a larger sector angle by steering.

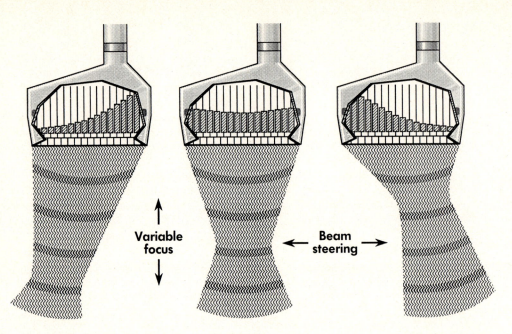

Variable focus

Beam steering

Fig. 11-28 Dynamically focused phased array beams involve even more complicated excitation schemes.

Review Questions: Chapter 11

1. Define or otherwise identify:
   a. Freeze frame
   b. Line density
   c. Frame rate
   d. Sector angle
   e. Segmental
   f. Steering
   g. Anulus
   h. Apodisation

2. List four distinct advantages of real-time imaging compared to compound B-scanning.

3. At approximately what frame rate does flicker disappear?

4. What is the trade off on frame rate, line density, and depth?

5. Compare the advantages and disadvantages of sector scanning.

6. If the frame rate is increased and the lines per frame are unchanged then

   a. speed of the US in soft tissue must change.
   b. the frequency must be increased.
   c. maximum imaging depth must be increased.
   d. maximum imaging depth must be decreased.

7. If the lines per frame are increased, while the imaging depth is unchanged, then
   a. the frame rate is increased.
   b. the number of shades of gray decrease.
   c. the frame rate is decreased.
   d. this cannot happen.

8. The number of lines per frame multiplied by the frame rate is equal to
   a. 1 M Byte.
   b. the frame refresh index.
   c. the pulse repetition frequency.
   d. the scan rate.

9. Discuss mechanical real-time scanners.

10. Discuss two methods of focusing a real-time scanner.

11. How does a phased array transducer work?

12 Image Processing and Display

The diagnostic ultrasound transducer generates ultrasound pulses by the piezoelectric effect, and as the beam propagates through the patient it encounters anatomical interfaces. At each tissue interface, some of the beam is reflected and some is transmitted according to the acoustic impedance differences. The reflected wave returns to the transducer where it deforms the crystal and produces an electric signal. This returning signal is generally weak and must be amplified and processed for display. This chapter will follow the ultrasound pulse from generation to display and discuss the instrumentation that makes clinical imaging possible.

The basic components of an ultrasound system that initiate and detect the ultrasound signal and present clinical information are diagrammed in Fig. 12-1. The pulser produces electric impulses that drive the piezoelectric crystal and tell the other components that the pulse has been sent. Echoes of varying intensities are acquired by the transducer acting as a receiver. After reception, the signals undergo amplification and processing and are then sent to the scan converter to be stored. From here the signal is transferred to the display from which it may be recorded on a hardcopy image. Each of the components will be examined individually.

PULSER

The pulser provides the rhythmic voltage spikes used to excite the crystal and generate the ultrasound pulses used for imaging. Typically the pulser provides rapid excitation pulses on the order of 200 to 600 volts at a rate of 1000 pulses per second (PRF = 1 kHz). The intensity of the ultrasound pulse produced depends on the peak voltage supplied by the pulser—the higher the peak excitation voltage, the greater the ultrasound intensity.

RECEIVER

The returning echoes are processed by the receiver. In the transducer, the receiving element is generally the same crystal used to transmit the ultrasound beam. The range of signals detectable by the receiving crystal is much smaller than the several hundred volts used by the transducer for pulse generation. Returning echo signals have a large range of amplitudes, usually from around 1 volt to as low as 1 μV. The **dynamic range** of the receiver or any device is the ratio of the smallest detectable input signal to the largest input signal that does not exhibit distortion. Dynamic range is measured in decibels (dB), and 120 dB is typically available in the receiver. This means that amplitude variations on the order of 1,000,000 to 1 can be acquired by the receiver without distortion. Thus the receiver has high **sensitivity** to extremely low level signals and to those signals that are more intense.

Even though the receiver has an extremely large dynamic range, other system components like the scan converter and display are limited to a much smaller dynamic range. Therefore the large spread of echo signals must be compressed into a narrower range. This process is termed **compression** and is performed in the receiver. Logarithmic compression is normally used since it allows very high and very low echo amplitudes to be compressed into ranges that can be seen on the display.

In addition to receiving returning echoes, the receiver incorporates several other functions: amplification, rectification or demodulation, and enveloping.

Amplification

The echo signals generated by the low Q imaging transducer must be **amplified** to about 10 V for display. Amplification, also referred to as **gain,** is normally expressed in decibels. The process of amplification is well known to those with a stereo system. When the volume control is turned up, the small electrical signals from the CD player are amplified and fed to the speakers. Chapter 5 (Eq. 5-2) dem-

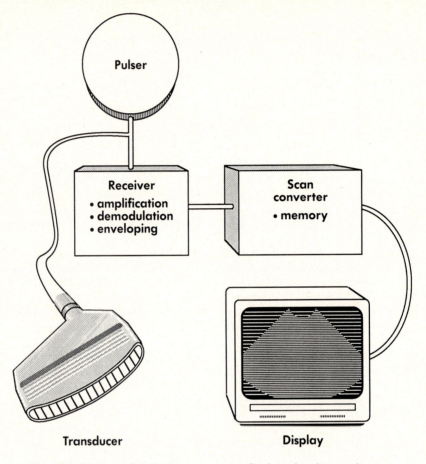

Fig. 12-1 Diagram of the basic components of pulse echo ultrasound system.

onstrated that a gain of signal amplitude from 1 to 100 is equivalent to a 40 dB gain. Modern clinical instruments have a maximum available gain of one million to one or 120 dB. Clinically, much of this dynamic range results from the attenuation of the ultrasound beam with depth. This attenuation can be compensated for by a selective form of amplification called time-gain compensation (TGC), also known as depth-gain compensation (DGC), or swept gain.

The ultrasound beam is attenuated at the rate of approximately 1 dB per cm per MHz in tissue. The time-gain compensation function amplifies the weaker echoes returning from deeper structures more than those returning from nearby reflectors. Thus, the later the echo returns to the transducer, the greater

the amplification it receives. Hence the name time-gain compensation. Fig. 12-2 illustrates attenuation of the beam with depth and restoration of the signal using the TGC process. Figure 12-2, *A* shows an idealized soft tissue equivalent phantom with interspersed thin-sheet reflective interfaces. The ultrasound beam reflected at each interface is represented by the returning arrows. Since the beam undergoes attenuation as it travels through the phantom, returning echoes will experience a decrease in amplitude as displayed in Fig. 12-2, *B*. TGC is used to compensate for this attenuation by selective amplification of signals from different depths. The result, shown in Fig. 12-2, *C*, is a uniform signal height from all depths. There are several portions of the

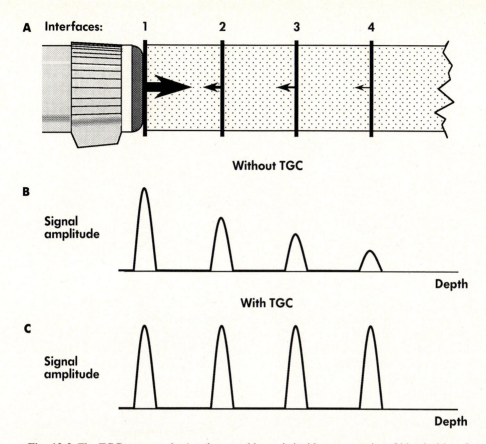

Fig. 12-2 The TGC process. **A,** An ultrasound beam is incident on a series of identical interfaces. **B,** The amplitude of the echo from each interface decreases with depth due to attenuation without time-gain compensation. **C,** Time-gain compensation is used to compensate for the loss of echo amplitude and thus restore each echo to the same amplitude.

TGC curve that are normally under the control of the sonographer.

Near gain. Near gain represents the amount of gain applied to the closest echoes. Since these echoes are normally very strong, minimal gain is required.

Delay control. Delay control regulates the time (depth) at which TGC begins to be applied. This function causes the closest echoes to be bypassed.

TGC slope. The TGC slope adjusts the degree of amplification added to echoes from increasing depths. Most units allow the user to design his own TGC versus range curve to compensate for the degree of attenuation in tissue. For example, higher frequency transducers require an increase in the TGC slope because of increased attenuation.

Far gain. Far gain is amplification applied only to distant echoes.

Enhancement. Enhancement amplifies echoes from a specific portion of the TGC curve, such as the mitral valve region, during cardiac evaluation.

Reject. Reject is available on some instruments to eliminate the smaller noise signals and unimportant echoes from the image.

A TGC curve illustrating several of these control functions is shown in Fig. 12-3.

Signal processing

Demodulation. The echo signals are first amplified and then subjected to demodulation. Demodulation involves rectification of an alternating voltage

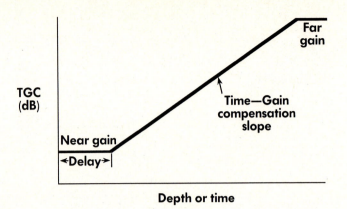

Fig. 12-3 Time-gain compensation curve produces echoes of equal amplitude.

waveform into a direct voltage waveform. It can be accomplished by either flipping up or by simply removing the negative portions of the echo signal. In either case, demodulation is the first step in decoding the amplitude information in the pulse to prepare it for gray scale display which cannot handle negative signals.

Enveloping. Enveloping, also termed smoothing, is a filtering process that is carried out after demodulation. The circuit traces an envelope around the demodulated signal, and the individual oscillations no longer appear. The resultant video envelope is a representation of echo amplitude with time. The signal must be enveloped to prevent the display from regarding the signal as resulting from several different echoes.

Manufacturers generally alter this signal shape further to make it more compatible with the display. Figure 12-4 illustrates three of the more common forms of processing. Leading edge detection (Fig. 12-4, *A*) places a spike at the beginning of the video signal envelope. This form is commonly used for bi-stable imaging. Peak detection (Fig. 12-4, *B*) substitutes a spike with the same height as the video peak for the video signal. With integration (Fig. 12-4, *C*), the spike height is proportional to the envelope area. Each of these methods replaces the video envelope with a single video spike. The series of steps involved in signal processing, known as **detection** of the signal, is summarized in Fig. 12-5.

SCAN CONVERSION

The scan converter accepts the output video signal from the receiver and stores the image data before it is sent to the TV display. The scan converter thus acts as the **memory** for the system and assigns echoes to a gray scale level. Storage is necessary because echo data is received at a rate that is very different from the rate required by the TV monitor. The scan converter is needed to make the incoming scan data compatible with the TV format. It also allows manipulation of the image into a preferred visual form and "freeze frame" for recording of the image.

There are two types of scan converters—analog and digital.

Analog scan converter

The analog scan converter, developed in the early 1970s, contains a two dimensional dielectric matrix with central elements capable of retaining an electric charge (Fig. 12-6). The target matrix typically has about 10^6 storage elements. Pulse-echo amplitude information is stored on the matrix elements by a scanning electron beam. The stored charge distribution pattern is a point by point representation of the echo amplitude pattern received from corresponding points within the patient. The greater the charge is on a dielectric element, the greater will be the echo amplitude received from the ultrasound beam. To read the stored information for video display, an electron beam is rapidly scanned over the

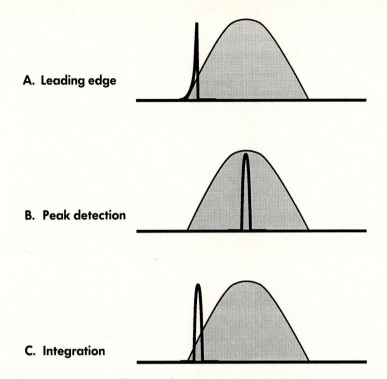

A. Leading edge

B. Peak detection

C. Integration

Fig. 12-4 Methods of final modification of enveloped signal. **A,** Leading edge detection—spike at beginning of the envelope is commonly used for bistable scanning. **B,** Peak detection—the maximum envelope height is represented by a spike. **C,** Integration—spike height is proportional to envelope area.

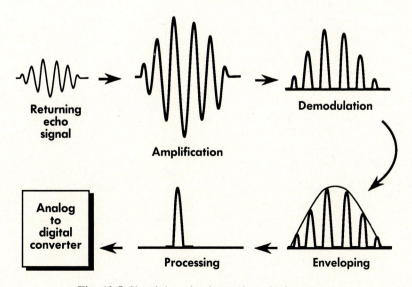

Returning
echo
signal

Amplification

Demodulation

Analog
to
digital
converter

Processing

Enveloping

Fig. 12-5 Signal detection in a pulse echo instrument.

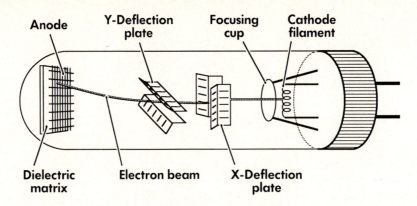

Fig. 12-6 Analog scan converter.

scan converter target. Analog scan converters, however, proved to have serious limitations, including instability. The image will deteriorate with time because of the drift of stored charge levels. Therefore, these units have been replaced in modern ultrasound equipment by digital technology.

Digital scan converter

Digital scan converters were developed to overcome the inherent instability problems found in analog systems. To use these devices, the echo signals must first be digitized. To understand digital devices, review the properties of binary numbers discussed in Chapter 2. In a binary system only two numbers, 0's and 1's called binary digits or bits, are used to represent all numbers. Fig. 12-7 demonstrates the conversion from binary to decimal numbers. As the numbers to be digitized become larger, more bits are usually required to form a multibit unit or word.

The number of bits in a word determines the number of values that can be used to represent echo amplitude signals. Table 12-1 displays information for word lengths up to eight bits.

Example: Convert 45 to a binary number. (Refer to Chapter 2.)

Answer: $45 = 32 + 0 + 8 + 4 + 0 + 1$
$= (1 \times 2^5) + (0 \times 2^4) + (1 \times 2^3) +$
$(1 \times 2^2) + (o \times 2^1) + (1 \times 2^0)$
$= 101101$

Fig. 12-8 illustrates the progress of a pulse echo through the imaging system. The echo signals are in analog format when they exit the receiver, but the

Table 12-1 Relationship between binary word length and the number of values represented

Word length (bits)	Possible number of values (2^n)	Range of decimal values	Maximum binary number
1	$2^1 = 2$	0-1	1
2	$2^2 = 4$	0-3	11
3	$2^3 = 8$	0-7	111
4	$2^4 = 16$	0-15	1111
5	$2^5 = 32$	0-31	11111
6	$2^6 = 64$	0-63	111111
7	$2^7 = 128$	0-127	1111111
8	$2^8 = 256$	0-255	11111111

digital scan converter can handle only discrete numbers that require a digital format. Therefore each echo signal is converted to a binary number in the analog-to-digital converter (ADC) before being stored in the digital memory. The digital scan converter stores the image data in a matrix array typically consisting of 512 × 512 elements called addresses or memory locations.

As illustrated in Fig. 12-9, each x-y location in the body has a unique address in memory. The echo amplitude from each location is converted to a binary number and then stored in the corresponding address in the scan converter. For simplicity, consider that the transducer is aimed so that only the first row of imaginary matrix elements within the patient are imaged. The first matrix element contains a highly re-

2 bit word:

$$1 \quad 1 = ?$$

$$
\begin{array}{cc}
1 & 1 \\
\times & \times \\
2^1 & 2^0 \\
\| & \| \\
2 & + & 1 & = 3
\end{array}
$$

3 bit word:

$$1 \quad 1 \quad 1 = ?$$

$$
\begin{array}{ccc}
1 & 1 & 1 \\
\times & \times & \times \\
2^2 & 2^1 & 2^0 \\
\| & \| & \| \\
4 & + & 2 & + & 1 & = 7
\end{array}
$$

$$0 \quad 0 \quad 0 = ?$$

$$
\begin{array}{ccc}
0 & 0 & 0 \\
\times & \times & \times \\
2^2 & 2^1 & 2^0 \\
\| & \| & \| \\
0 & + & 0 & + & 0 & = 0
\end{array}
$$

5 bit word:

$$1 \quad 1 \quad 1 \quad 1 \quad 1 = ?$$

$$
\begin{array}{ccccc}
1 & 1 & 1 & 1 & 1 \\
\times & \times & \times & \times & \times \\
2^4 & 2^3 & 2^2 & 2^1 & 2^0 \\
\| & \| & \| & \| & \| \\
16 & + & 8 & + & 4 & + & 2 & + & 1 & = 31
\end{array}
$$

$$0 \quad 1 \quad 0 \quad 1 \quad 0 = ?$$

$$
\begin{array}{ccccc}
0 & 1 & 0 & 1 & 0 \\
\times & \times & \times & \times & \times \\
2^4 & 2^3 & 2^2 & 2^1 & 2^0 \\
\| & \| & \| & \| & \| \\
0 & + & 8 & + & 0 & + & 2 & + & 0 & = 10
\end{array}
$$

Fig. 12-7 Conversion of binary numbers of different word lengths to decimal numbers.

flective interface that results in a strong echo signal. This signal is processed and then converted in the analog-to-digital converter to 111111 in binary (63 in decimal). Note that the scan converter memory requires a depth of six bits to store the six 1's. The second matrix element in row one in the patient, which contains a weakly reflecting interface, is assigned the binary number 001000 (8 in decimal). Each address in the digital scan converter is filled in a similiar manner, with converted echo signals from each location in the body.

The number of bits in the memory-bit depth, sometimes called memory resolution, determines how many shades of gray the image may contain. A single matrix element can store only a one or zero in each memory location. This would allow only black and white or **bi-stable** imaging, since locations with any echo signal could only be assigned a 1 (white), and those with no echo signal would be assigned 0 (black). Incidently, a great deal of the early pioneering clinical work in diagnostic ultrasound was done with bi-stable scanning.

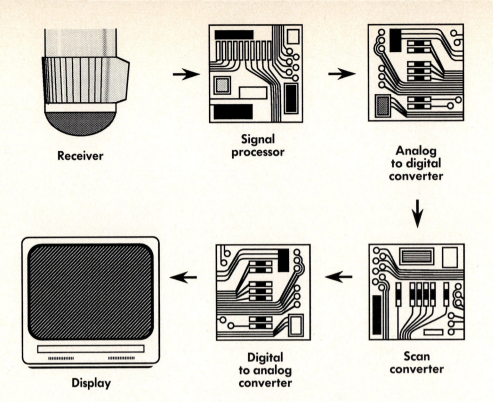

Fig. 12-8 Components of an ultrasound imager.

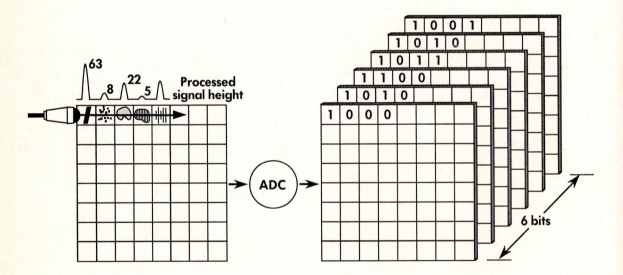

Fig. 12-9 Transfer of information from the receiver to the corresponding location in the digital scan converter. The echo amplitude signals are converted to binary numbers in analog to digital converter and sorted in 6-bit deep memory. Storage of four echo strengths—111111 (63), 001000 (8), 010110 (22), and 000101 (5)—are illustrated.

Gray scale imaging requires more matrices as shown in the simplistic example in Fig. 12-9. The scan converter matrix shown is only 8×8 with a depth of six bits. Thus only binary numbers representing decimal values ranging from 0 to 63 can be stored in each memory location. Most ultrasound equipment manufactured today uses six-or eight-bit deep memories that provide 64 (2^6) or 256 (2^8) shades of gray at each tiny picture element or **pixel** location on the display. The binary numbers are read out of memory, reconverted to analog signals in the digital-to-analog converter (DAC), and then sent to the TV monitor for display.

PREPROCESSING AND POSTPROCESSING

Preprocessing occurs before echo amplitudes are stored in the scan converter. Preprocessing is the manipulation of the digitized echo signals before storage. For many systems, preprocessing is not controlled by the operator. Other systems offer the sonographer a choice of preprocessing curves with different functions. For example, a linear curve may assign echo strength levels equally to the available digital values. Another curve may enhance the differentiation of low amplitude signals by spreading these low dB echos over most of the range of the digital numbers available. Typical examples of preprocessing curves are shown in Fig. 12-10.

Postprocessing is the assignment of gray levels in the image to the numbers stored in the digital scan converter. The number of gray shades possible depends on the bit depth of the memory. The six-bit dynamic range illustrated in Fig. 12-9 has potential for 2^6 or 64 shades of gray. Postprocessing does not alter digitized values; it is the assignment of these values to different brightness levels on the TV monitor to optimize the presentation of information. Fig. 12-11 demonstrates the effect of four different postprocessing techniques on a gray scale image.

Even 64 shades of gray may seem like overkill since the human eye can distinguish only about 16 shades of gray. However, with additional postprocessing functions, such as level adjustment and windowing, an even greater dynamic range can be usefully applied.

DISPLAY DEVICES

The device used to display the ultrasound image is a cathode ray tube (CRT). CRTs come in several different forms: oscilloscopes, CRT storage displays, and television monitors (TV). Today, the most

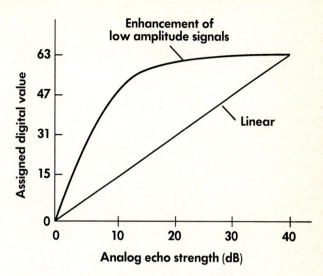

Fig. 12-10 Typical preprocessing curves that assign the analog signal level to a digital value. The linear curve assigns values uniformly while the top curve enhances the separation of signals below 20 dB.

commonly used display device is a TV because it provides excellent gray scale rendition at low cost. TV monitors also allow control of the brightness and contrast, the use of color, and a zoom feature that uses only part of the stored data to generate the TV image.

The function of each CRT device is similar. As shown in Fig. 12-12, *A*, the CRT contains a large evacuated glass tube with a high potential difference from cathode to anode. Monochrome CRTs operate at about 18,000 volts while color CRTs require 32,000 volts. Electrons are boiled off the "electron gun" filament by thermionic emission. The focusing cup keeps the electrons in a narrow beam. The intensity of the electron beam is regulated by a control grid, which is attached to the electron gun. The video signal received by the picture tube is modulated; that is, its magnitude is directly proportional to the echo amplitude stored in the scan converter.

Inside the front face of the glass tube is a coating of phosphorescent material that produces light when struck by electrons. As the intensity of the electron beam increases so does the brightness of the light from the phosphor. The glass envelope is usually backed by a thin layer of aluminum that transmits

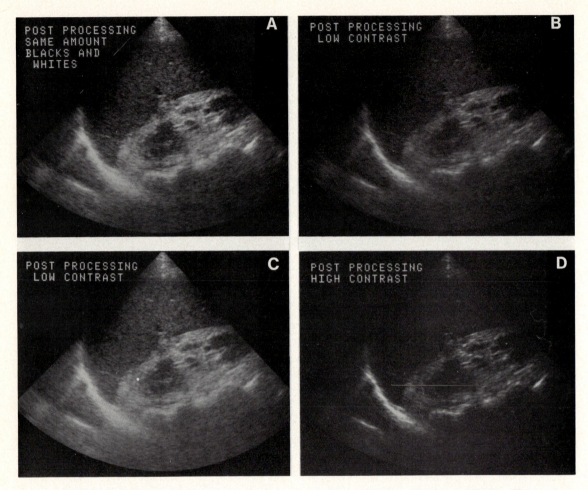

Fig. 12-11 Postprocessing assigns numbers in memory to a brightness value on the display. The four scans shown illustrate the effect of different postprocessing assignments on the gray scale image. **A,** Linear curve. **B,** Low contrast curve. **C,** Low contrast curve (slightly higher contrast than B). **D,** High contrast.

the electron beam, but reflects the light. The electron beam is directed across the tube face by the deflecting coils in the rapid left-to-right or **raster** scan sequence shown in Fig. 12-12, *B*.

The variable intensity electron beam begins in the upper-left corner of the CRT screen and moves to the upper-right corner, tracing a line of varying intensity light as it moves. This is called an active trace. The electron beam is then blanked, or turned off, and it returns to the left side of the screen as shown. This is called the horizontal retrace. There is then a series of active traces followed by horizontal

retraces until the electron beam is at the bottom of the screen. This is much like the action of a typist who types a line of information (the active trace), returns the carriage (the horizontal retrace), and continues this sequence to the bottom of the page. Whereas the typist completes a page, the electron beam completes a television field.

The similarity stops there, however, because the typist would continue on another page. The electron beam is blanked again and undergoes a vertical retrace to the top of the screen. It now describes a second television field, the same as the first, except

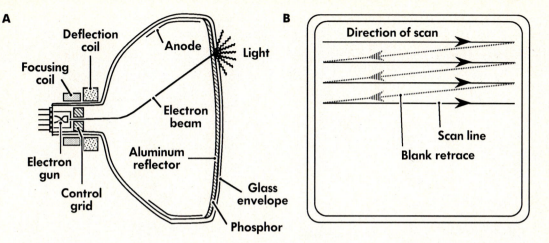

Fig. 12-12 A, A picture tube (CRT) and its principal parts. **B,** Illustration of raster scan pattern.

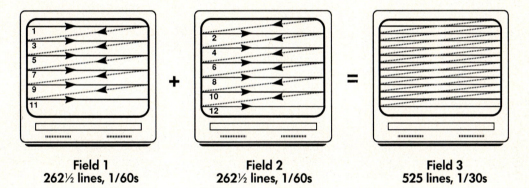

Field 1
262½ lines, 1/60s

Field 2
262½ lines, 1/60s

Field 3
525 lines, 1/30s

Fig. 12-13 A video frame is formed from a raster pattern of two interlaced video fields.

that each active trace lies between two adjacent active traces of the first field. This movement of the electron beam is termed interlace, and two interlaced television fields form one television frame (Fig. 12-13).

In the United States our power is supplied at 60 Hz; therefore there are 60 television fields per second and 30 television frames per second. This is fortuitous because the flickering of home movies, shown at 16 frames per second, or of old-time movies—the "flicks"—does not appear on the television image. Flickering is not detectable by the human eye at rates above 20 frames per second. At a frame rate of 30 per second, each frame is 33 ms long.

Standard broadcast and closed circuit TV are called 525-line systems because they have 525 lines of active trace per frame. Actually there are only about 490 lines per frame because of the time required for retracing.

To display the echo information stored in memory, the memory readout is synchronized with the TV's electron beam scan. The position of each matrix element in the scan converter is identical to the position on the monitor. Echo amplitudes are displayed with different brightness levels according to the stored number and postprocessing technique selected. A large echo signal will produce a large-stored value in a scan converter matrix element and

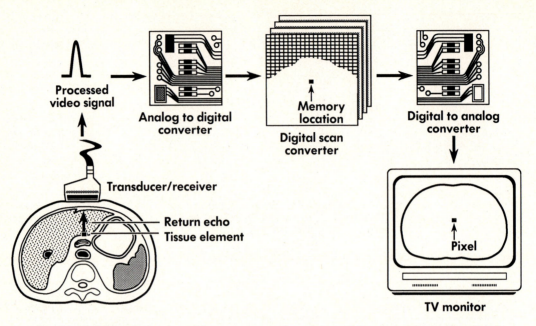

Fig. 12-14 Schematic drawing of the components of an ultrasound imaging system with a digital scan converter. An echo signal returning from an interface in the patient is processed into a voltage spike of appropriate amplitude. This analog signal is then converted into a binary number in the ADC and sorted in the appropriate matrix location within the digital scan converter. For display on a TV monitor, the signal is read out of memory and converted back to analog in the DAC.

thus a bright spot on the monitor. To form the image, the electron beam "paints" a series of black, white, or gray scale dots as it scans across the TV monitor face. The numbers that are read out of memory tell the electron beam whether to apply a longer burst (white dot) or shorter burst (darker dot) to each location along the scan. The resulting mosaic is the gray-scale image. Fig. 12-14 reviews the sequence of steps normally required in the production of the ultrasound image ultimately displayed on the TV monitor.

IMAGE RECORDING
Polaroid images

It is important that the images from the CRT monitor be recorded or reproduced accurately for later inspection and storage in the patient's file. Although the recent trend in ultrasound image recording has been toward multi-image recording devices, "instant" single-image photographic techniques are still in use. Some facilities still use Polaroid cameras to record the ultrasound image directly from the mon-

itor. To optimize Polaroid images, it is essential to maintain proper photographic technique. To test this, a gray-scale bar generated by most ultrasound equipment should be photographed and evaluated for appropriate image quality.

The recommended development time is important. Allow 3 to 5 minutes before peeling the Polaroid print from the negative. If the development time is too long, the images will rapidly degrade.

Excessive humidity will also degrade the Polaroid image. Since the foil wrapper on the Polaroid film pack guards the film against excessive humidity, it should not be removed until the film is ready for use.

Multiformat cameras and laser printers

Multi-image film cameras, which use 8 × 10 inch x-ray film, have been widely used for recording ultrasound images (Fig. 12-15). More recently, multi-input imagers and laser printers have become available for recording images. These devices can be docked directly to a processor to provide a complete

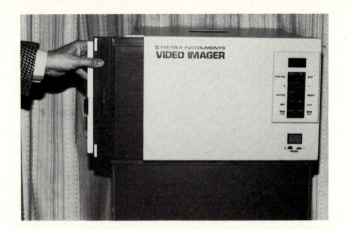

Fig. 12-15 Multi-format camera used in ultrasound image recording. (Courtesy Agfa Matrix).

cassetteless, film-handling, imaging, and processing system. Many sonographers, however, have very limited knowledge of how the devices function. Time spent with the operator's manual and service personnel will improve one's technical competence and reduce later frustrations. Upon installation, the imager should be thoroughly acceptance-tested by a qualified person, and the test films retained as documentation. The acceptance-test images, made with the unit in good working order, can be used later to compare with questionable or unacceptable images.

Since hardcopy imagers require the use of a film processor, it is essential, for quality control, that correct solution temperatures and chemistry are used. Small deviations in developer temperatures, developer contamination, or improper replenishment rates can severely degrade image quality. Every automatic processor used in ultrasound should be subjected to quality control evaluations. For those imaging departments that do not have a processor-quality-control program, an x-ray film sales representative can assist in setting up such a program.

Other recording modalities

Thermal processors have recently been introduced into the video copy market. These devices contain an internal memory capable of using the digital echo amplitude signals stored in the scan converter to produce a gray scale image printed on thermal paper.

Video cassette recorders (VCR) have found application in cardiology and some diagnostic imaging facilities. The video tape contains a layer of ferrom-

agnetic material such as a metal oxide. Recording the video image is accomplished by moving the tape past an electromagnet called the recording head that changes the alignment of the magnetic particles in response to the applied field. The field strength varies directly as the echo strength from each pixel location.

A large number of VCR images can be stored on inexpensive cassette tape and viewed in cine or freeze-frame mode. VCR equipment requires the periodic use of a special cassette to clean the recording and playback heads.

A quality assurance program for image-recording devices should be incorporated into the overall department program outlined in Chapter 15.

Review Questions: Chapter 12

1. Define or otherwise identify:
   a. Receiver
   b. Pulser
   c. Dynamic range
   d. Sensitivity
   e. Amplification
   f. Demodulation
   g. Enveloping
   h. Scan converter
   i. Pixel
   j. Memory bit depth
   k. Cathode ray tube
   l. Interlacing

2. Place the following in the correct sequence.
   a. Display
   b. Reception
   c. Processing
   d. Recording
   e. Scan conversion
   f. Amplification
   g. Transmitting pulse
   h. Reflected echo

3. Briefly explain logarithmic compression.

4. Discuss the TGC controls that are normally adjustable by the sonographer.

5. What does the scan converter do?

6. Compare the function of the analog and digital scan converter. What were the limitations of the analog scan converter that lead to its replacement?

7. Describe the feature of an ultrasound system that is used to compensate for tissue attenuation.

8. Describe the relationship between bits and bytes. How are these used in ultrasound?

9. Illustrate how the gray scale is represented in a digital image.

10. Trace the ultrasound pulse from reflection to display. How does the instrument ensure that the location of an object presented on the display is the same as the actual location in the patient?

11. Briefly discuss three methods of recording images.

13 Doppler Ultrasound

The Doppler effect was first described by Austrian physicist Christian Doppler in 1842. Doppler found that there is an apparent change in the frequency of a wave if there is relative motion between the wave source and the observer. He conducted a simple experiment with a bugler on a moving train and another bugler on the station platform. Both buglers were playing the same musical note but the frequency heard was not the same. He found that if the bugler on the train was moving toward the station, the sound from the moving bugler heard by an observer on the platform was at a higher frequency than the sound from the stationary bugler. Conversely, when the train had passed and the source and observer were moving apart, the observer heard a lower frequency from the moving bugler than the stationary bugler. This was an historic experiment.

A familiar example of the Doppler effect is that of a passing train (Fig. 13-1). If the train's whistle has a frequency of 3500 Hz, the wavelength will be about 10 cm. Thus, each pulse will travel about 10 cm before the next pulse is emitted. When the train is approaching, the peaks will tend to bunch up so that the wavelengths that reach the observer are less than 10 cm. Since the wavelength is smaller, the frequency is higher, and the observer will hear a pitch higher than the orginal 3500 Hz. The velocity of the individual sound waves in air do not change. It is the decrease in wave spacing caused by the motion of the source that produces the increase in frequency. After the train passes, the waves are spread out by more than 10 cm. This longer wavelength means that the frequency heard is lower than 3500 Hz. Have you heard the abrupt change in whistle frequency that occurs when a train speeds past you at a crossing?

The Doppler effect has also been used in astronomy to determine the relative motion of stars with respect to earth. Stars moving away from the earth produce light that appears to be shifted toward the red or longer wavelength region of the spectrum. The light from stars moving toward the earth has higher frequency and appears more blue. Although not standardized, color-flow imaging sometimes employs the same color rendition for motion toward and away from the transducer. It is important to remember throughout this discussion that Doppler ultrasound pertains only to moving objects or interfaces.

DOPPLER MATHEMATICS
The Doppler equation

In diagnostic ultrasound, the Doppler effect is used to non-invasively detect blood flow and the motion of body structures as shown in Fig. 13-2. When the ultrasound beam is reflected from a moving object, the frequency of the reflected beam will be different from the initial source frequency. Doppler instrumentation presently includes continuous wave (CW), pulsed, and color flow. Each type of instrument relies on the mathematical formulation of the Doppler principle to describe the velocity and, in most cases, the direction of motion of interfaces within the body.

Definition: Doppler Shift Frequency is the difference between the frequency reflected from the moving interface (f_R) and the transmitted frequency (f_T):

Symbol: f_D

Equation: $$f_D = f_R - f_T \qquad \text{(13-1)}$$

Example: What is the Doppler shift frequency if the frequency transmitted by an ultrasound transducer is 5 MHz and that reflected is 5.01 MHz?

Answer:
$$f_D = f_R - f_T$$
$$f_D = 5.01 \text{ MHz} - 5.00 \text{ MHz}$$
$$f_D = 0.01 \text{ MHz} = 10,000 \text{ Hz}$$

Fig. 13-1 Doppler effect describes the apparent change in frequency of emitted sound when either the source or the receiver is in motion. As a train passes, the pitch of its whistle changes from high to low.

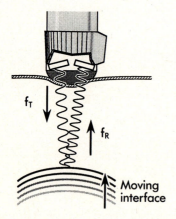

Fig. 13-2 Example of change in frequency (Doppler shift) of CW ultrasound beam reflected from interface moving toward the transducer. The reflected waves are compressed causing the received frequency to be greater than that of the source.

This example illustrates two important aspects of Doppler ultrasound. First, since the reflected frequency is higher than that transmitted, the reflector is moving toward the transducer. If the interface were moving away from the transducer, f_D would be negative since f_R would be less than f_T. Second, the Doppler shift frequency that results is in the audible range. This is a common occurrence for motion encountered clinically.

The Doppler shift frequency can also be written in terms of the velocity of ultrasound in the medium (v) and the velocity of the interface (u):

$$f_D = f_T (2u/v) \qquad\qquad \textbf{(13-2)}$$

The value of u is assigned a plus sign if the interface is moving toward the source and a negative sign if the motion is away from the source. Eq. 13-2 indicates that the Doppler shift frequency will be greater for higher transducer frequencies (f_T) and faster moving structures (u).

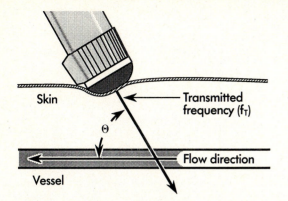

Fig. 13-3 The Doppler shift frequency depends on the Doppler angle, θ. Normally Doppler angles between 30° and 60° are employed.

Table 13-1 Values of Cos θ for representative Doppler angles

Doppler angle (θ)	Cos θ
0°	1.000
15°	0.966
30°	0.866
45°	0.707
60°	0.500
75°	0.259
90°	0.000

Example: If a 2 MHz transducer scans a soft tissue interface that has a velocity of 20 cm s^{-1} toward the detector, (a) what is the Doppler shift frequency, and (b) what frequency will be seen by the detector?

Answer: a. $f_D = f_T (2u/v)$
$f_D = (2 \text{ MHz})(2)(0.2 \text{ m s}^{-1})/1540$ m s^{-1}
$f_D = (2 \times 10^6 \text{ s}^{-1})(0.4 \text{ m s}^{-1})/$ 1540 m s^{-1}
$f_D = 519 \text{ s}^{-1}$
$f_D = 519 \text{ Hz}$

b.   Since the motion of the interface is toward the detector, solve Eq. 13-1 for the reflected frequency:
$f_D = f_R - f_T$
$f_R = f_D + f_T$
$f_R = 2,000,000 \text{ Hz} + 519 \text{ Hz}$
$f_R = 2,000,519 \text{ Hz}$
$f_R = 2.000519 \text{ MHz}$

In this example, the frequency of ultrasound reflected from the moving interface is only slightly higher than the original 2 MHz frequency transmitted. If the Doppler shift frequency is known, Eq. 13-2 can be used to compute the velocity and direction of the moving interface. This is done by solving the Doppler equation for the interface velocity:

$$u = vf_D/2f_T \qquad (13\text{-}3)$$

Example: Find the velocity of an interface moving toward the 5 MHz transducer, if the Doppler shift frequency is 3000 Hz.

Answer: $u = v f_D/2f_T$
$u = (1540 \text{ m s}^{-1})(3 \times 10^3 \text{ s}^{-1})/2(5 \times 10^6 \text{ s}^{-1})$
$u = 0.46 \text{ m s}^{-1} = 46 \text{ cm s}^{-1}$

The Doppler angle

Doppler ultrasound is often used to evaluate blood flow. The red blood cells act as scattering centers. Unlike static B-mode or real-time, which give the best images when the beam has perpendicular incidence, the Doppler shift signal is the largest when blood flow is directly toward or away from the transducer. Eq. 13-2 is valid only for such motion along the axis of the ultrasound beam as shown in Fig. 13-2. Normally, parallel transducer orientation is not possible. Usually the beam is at an angle θ with respect to the vessel as depicted in Fig. 13-3.

Definition: Doppler angle is the angle that the ultrasound beam makes with the direction of flow.

Symbol: θ (Greek letter "Theta")

Equation: Angular dependence is incorporated into the Doppler equation using the cosine function.

$$f_D = f_T (2u/v) \text{ Cos } \theta \qquad (13\text{-}4)$$

The cosine is a trigonometric function varying between 1 and 0 as θ varies from 0° to 90°. Table 13-1 provides values of Cos θ for a number of Doppler angles.

Fig. 13-4 illustrates the effect of incident beam angle on Doppler shift frequency. A 5 MHz trans-

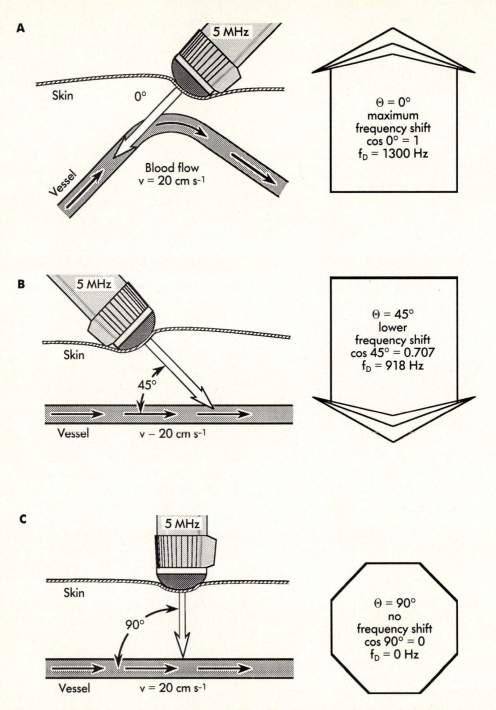

Fig. 13-4 The magnitude of the Doppler shift frequency depends on the Doppler angle. **A,** The maximum Doppler shift frequency occurs when motion is directly toward (or away from) the transducer. **B,** At 45° beam angulation the Doppler shift frequency is reduced by 30%. **C,** If the beam is perpendicular to the vessel, in theory no Doppler shift is detected since cos 90° = 0.

Table 13-2 Doppler shift frequencies for various Doppler angles and interface speeds

Original frequency (MHz)	Doppler angle (θ)	Interface velocity (cm s⁻¹)	Reflected frequency (MHz)	Doppler shift (Hz)
2	0°	50	2.00130	1300
2	60°	50	2.00065	650
2	0°	200	2.00520	5200
5	0°	50	5.00325	3250
5	0°	200	5.01300	13,000
5	90°	200	5.00000	0

ducer is directed at a vessel containing blood flowing through it at the rate of 20 cm s⁻¹. At 0° incidence (Fig. 13-4, *A*), the value of the cosine is one and the Doppler shift frequency determined from Equation 13-4 is 1300 Hz. If the Dopper angle is 45° (Fig. 13-4, *B*), the frequency shift is reduced by more than 30% to 918 Hz. Even though the Doppler equation indicates that $f_D = 0$ when $\theta = 90°$ (Fig. 13-4, *C*), there will be in practice some shift because of beam divergence. In general, as seen from the cosine values in Table 13-1, the smaller the value of θ, the greater the Doppler shift frequency.

Example: A 2 MHz transducer scans an interface moving toward the detector at 20 cm s⁻¹. What is the frequency shift if the Doppler angle is 60°? Compare answer with previous example.

Answer:
$$f_D = f_T (2u/v) \cos \theta$$
$$= 2 \times 10^6 \text{ s}^{-1} (2)(0.2 \text{ m s}^{-1}/1540 \text{ m s}^{-1}) \cos 60°$$
$$= 519 (.50) \text{ s}^{-1}$$
$$f_D = 259.5 \text{ Hz}$$

Thus, angling the transducer at 60° reduces the Doppler shift frequency by one half of the 519 Hz calculated in the previous example.

In practice, manufacturers usually require that the transducer be positioned at a 30° to 60° angle with respect to the flow direction to obtain a diagnostically useful Doppler signal. Angles larger than 60° result in too little frequency shift, and those less than 30° usually produce high beam attenuation because of the resultant longer path lengths. The precise angle used is not critical if one is interested only in determining the presence of flow.

For absolute velocity measurements, accurate knowledge of the Doppler angle is essential. Most state of the art Doppler instruments provide an angle cursor that marks the path of the beam on the freeze-frame gray scale monitor image. This provides the angle θ, the velocity in tissue is programmed in, and the Doppler shift frequency is computed by the system. With this information the interface velocity can be accurately determined using Eq. 13-4. However, if the Doppler angle is not measured and incorporated into the calculation, the velocity computed by the machine is meaningless and must not be used.

Thus the Doppler shift frequency may be stated either in kHz or converted to velocity by use of Eq. 13-4. Expressing the Doppler shift in kHz implies that one does not know the angle, or that it cannot be accurately measured. Table 13-2 displays some typical values of Doppler shift for various Doppler angles and interface velocities.

DOPPLER MODES
Continuous wave Doppler

Continuous wave (CW) Doppler instruments continuously transmit and receive an ultrasound signal. There are no pulses or pulse echos. The CW Doppler transducer (Fig. 13-5) contains two crystal elements—one for transmitting the continous signal, the other for receiving the echo. The crystals are usually slightly angled toward one another to produce an overlapping region of maximum transducer sensitivity where the beam patterns cross. The CW Doppler unit will process the echoes from any moving interface within this region.

Typical frequencies emitted by the transmitting crystal range from 2 to 10 MHz, depending on the application. Since higher frequency transducers result in greater beam attenuation, they are used for superficial vessels and structures. However, it is important to use a transducer with the highest possible frequency for any given application because the intensity of the echo signal scattered from blood cells increases dramatically with increasing frequency.

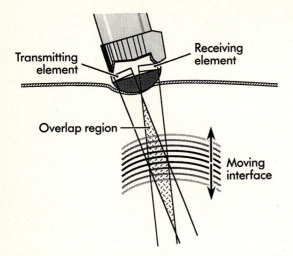

Fig. 13-5 Representation of a continuous wave Doppler transducer. CW Doppler systems detect all moving objects in the beam overlap region (cross-hatched area).

Since the Doppler shift depends on a constant, continuous source frequency, CW Doppler transducers employ high Q crystals. The transmission crystal is continuously excited by a changing electric field that oscillates at constant frequency. Unlike transducers used in pulse-echo imaging, no damping is applied. Air is usually the backing material, and that ensures maximum reflection from the back surface of the transmitting crystal. This large acoustic impedance difference allows nearly 100% of the beam to be reflected back into the patient.

The second crystal in the CW Doppler probe receives the returning echoes from both stationary and moving interfaces in the overlap region shown in Fig. 13-5. The signal received from non-moving structures contains echoes equal to the initial transmitted frequency, while moving interfaces produce Doppler shifted frequencies. Since the only signal of interest is that from the moving structures, the instrument subtracts the transmitted frequency from the reflected frequencies to generate the Doppler signal. This process, known as demodulation, is similar to the method used to separate the music from the carrier radio waves in an FM radio.

Doppler shift frequencies usually range from 200 Hz to 15 KHz and thus are in the audible range. An audioamplifier and speaker can be used to "listen to"

the moving interface. A familiar application of CW Doppler is the audible monitoring of the fetal heart by a system called an ultrasonic stethoscope. The emission of higher frequency sound indicates a large Doppler shift and therefore faster moving regions. Low frequencies indicate that the beam is picking up structures moving with lower velocity. The frequency shift information can also be displayed on a strip-chart recorder to help distinguish the motion of several moving objects in the overlap region.

Simple CW instruments can detect the motion of reflectors in the sensitive region but not the direction of the motion. These are termed non-directional systems. To determine if the flow or interface is moving toward or away from the transducer, a rather complex system called a **quadrature phase detector** is required. This is an electronic circuit that distinguishes direction by examining the phase shifts of the returning echoes. Output devices for such bidirectional CW Doppler systems include stereo speakers (one for each direction of motion) and chart recorders. A representative system is shown in Fig. 13-6.

Some modern CW Doppler systems include freeze-frame real-time images that are updated several times each second to assist the sonographer in identifying the structures being subjected to Doppler analysis. The gray-scale image is also useful in determining the Doppler angle, which is required to accurately determine velocity. The real-time image is generally updated several times per second to aid in positioning the Doppler beam.

The primary disadvantage of CW Doppler systems is that all moving interfaces within the sensitive region of the transducer will be detected. Moving structures at different depths are sampled simultaneously. This lack of depth resolution makes it difficult to determine exactly which site is being recorded. Thus other Doppler instruments have been developed to overcome this limitation.

Pulsed Doppler

Pulsed Doppler was designed to overcome the lack of range resolution inherent in CW Doppler by allowing only the Doppler signal from a selected depth to be detected and processed. By pulsing the beam in a manner similar to a pulse-echo imaging system, the pulsed Doppler instrument can obtain both position and velocity information. Like pulse-echo imaging transducers, the pulsed Doppler transducer uses the same crystal for transmitting and re-

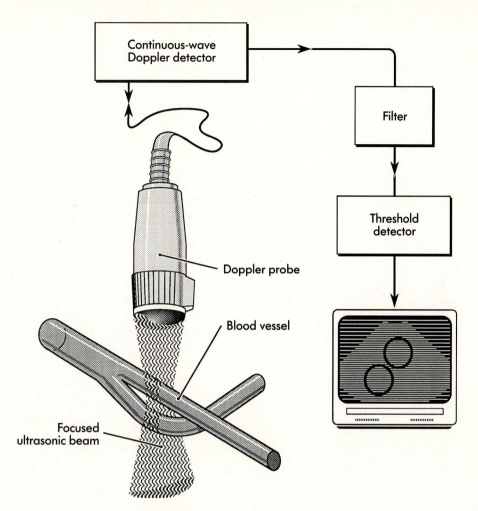

Fig. 13-6 Bidirectional continuous wave Doppler system.

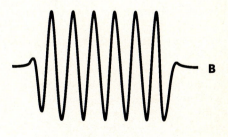

Pulse echo imaging

Pulsed Doppler

Fig. 13-7 A comparison of typical pulses used for imaging and for Doppler analysis. **A,** Pulses used for imaging are short in order to provide good axial resolution and they contain a wide range of frequencies. **B,** Pulses used for Doppler are of longer duration and narrow bandwidth.

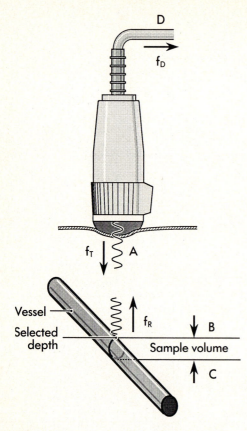

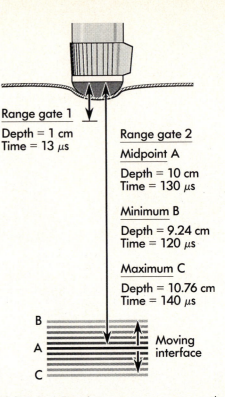

Fig. 13-8 Sequence of events in pulsed Doppler technique. **A,** Transducer sends Doppler pulse then is turned off. **B,** Receiver turned on after preset time so signal processing begins at selected depth only. **C,** Receiver turned off after short time—determines sample volume. **D,** Frequency shifted echo from sample volume is processed and displayed.

Fig. 13-9 Pulsed Doppler systems use range gating to selectively receive velocity information from a small region along the ultrasound beam path. The transducer is allowed to listen for return frequency shifted echoes only after a fixed time comparable to the depth that the operator wishes to examine. Range gate 1 detects signals only at a depth of 1 cm. Range gate 2 detects signals only from a depth of from 9.24 cm to 10.76 cm since the unit will listen for frequency shifts only from pulses that have round trip travel times between 120 and 140 μs.

ceiving. However, unlike imaging ultrasound pulses, Doppler ultrasound pulses have a narrow bandwidth and contain a larger number of cycles per pulse— typically from 5 to 20 cycles. A comparison of pulses used for imaging and for Doppler is illustrated in Fig. 13-7.

The pulses returning from moving interfaces are Doppler-shifted to higher or lower frequencies according to Eq. 13-4. Longer pulse lengths provide greater sensitivity because the additional cycles can be used for more accurate determination of the Doppler shift. Manufacturers now incorporate pulsed Doppler into many of their real-time imaging systems. Such duplex Doppler systems enable the operator to evelute the flow dynamics of any isolated area on the B-scan.

In all types of pulsed Doppler systems, depth in-

formation is derived by allowing the crystal to "listen" for a limited time following each transmitted pulse. This process is called **range gating.** The beginning of the "listening window," which is set by the operator, corresponds to a specified depth in tissue. The sample volume or gate width, from which the Doppler signal is received, is related to the length of time the crystal is allowed to listen to the returning echoes. A small sample region can be chosen by allowing the transducer to listen for only a short time after it is activated. This sequence of events required for pulsed Doppler is depicted in Fig. 13-8.

If the range gate is set at 13 μs (round trip), as Fig. 13-9 shows, only the motion of an interface at 1 cm from the transducer could be analyzed. This distance is readily verified by use of the range equation introduced in Chapter 11 (see Review Question

Apparent direction of rotation

Fig. 13-10 The apparent backward rotation of wagon wheels in movies is due to aliasing. Aliasing occurs when the sampling rate is not sufficient to adequately record the motion. Here the wagon wheels are turning faster than the camera frame rate. In pulsed Doppler systems aliasing may be manifested as inaccurate velocity or direction of flow.

12). Setting the range gate to include only echoes received from 120 to 140 μs (after transmission) would allow imaging of flow within a vessel with a 1.52 cm cross section, centered 10 cm from the transducer. Range gating allows determination of velocity information from specific depths without interference from Doppler signals from other regions. Multiple gates may be used to obtain flow information from several depths.

Transducers designed for pulsed Doppler applications include both mechanical and electronic models. Mechanical sector systems alternate between pulse-echo and Doppler modes. These single element systems use a mechanically focused crystal that sweeps out a sector of anatomy through an angle selected by the sonographer. However, because these systems use fixed focal length mechanical focusing, different transducers are required for scanning at different depths. Selecting an annular array transducer, which provides focusing throughout a larger range of depths, alleviates this problem.

Phased-array transducers electronically sweep and focus at a range of depths. Sophisticated electronics in the color flow system permit simultaneous display of color Doppler and gray-scale images. With these systems, the return echoes are evaluated for frequency shift and phase information, in addition to the pulse-echo amplitude and spatial data. This velocity and phase information enable any moving interface to be displayed in color.

Aliasing

Pulsed Doppler systems are limited in their ability to measure blood flow rates found in very deep vessels. Because of the time required to listen for echoes from deeper structures, most pulsed Doppler systems decrease the pulse repetition frequency as the depth is increased. When the number of pulses per second (pulse repetition frequency) is less than twice the Doppler shift frequency of the sample region, the Doppler shift signal is distorted by a phenomenon called aliasing. It is the result of insufficient sampling caused by inadequate pulse repetition frequency and results in false low frequencies in the Doppler signal. Aliasing is familiar to us from Western movies (Fig. 13-10) in which stagecoach wheel spokes appear to turn in the wrong direction. The frame rate of the film is too slow (inadequate sampling) to record the true speed and direction of the spinning wheels. The maximum Doppler shift frequency that can be accurately sampled is known as the Nyquist limit and is one half of the pulse repetition frequency.

Example: What is the maximum Doppler shift that can be sampled without aliasing if the pulse repetition frequency of the system is 5 KHz?

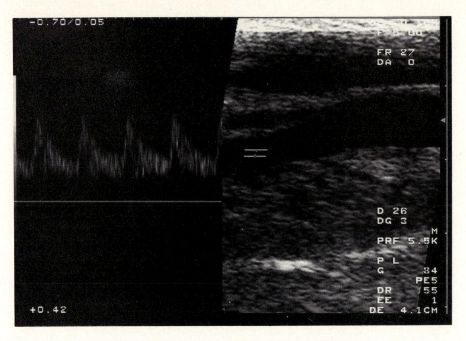

Fig. 13-11 A Doppler spectrum of the carotid artery.

Answer: The maximum Doppler shift is PRF/2 = 2.5 KHz.

Increasing the pulse repetition frequency will increase the sampling rate and decrease aliasing. However, the pulse repetition frequency cannot be increased indefinitely because the pulse might have time to travel to the depth of interest and return to the transducer before the next pulse is emitted. As a rule, the deeper in tissue, the lower the velocities required to produce aliasing. Some pulse Doppler units use automatic adjustment of the PRF based on the sample volume depth, while other units allow manual control of the PRF. Several systems are also equipped with CW Doppler that may be used when the PRF cannot be increased enough to prevent aliasing when fast flow occurs within deep vessels.

Spectral analysis

When the Doppler beam is targeted on a particular blood vessel, passing red blood cells will scatter the ultrasound beam. The fraction of the signal scattered back into the transducer will contain a variety of frequencies because red cells move at different velocities through the vessel. Current Doppler units contain sophisticated instrumentation that can analyze and provide a gray-scale display of the frequency (or velocity) spectrum of these blood cells as they move through the vessel. These instruments manipulate the complex Doppler echo and process it into the spectral components shown in Fig. 13-11, using a device called the Fast Fourier Transform. This process, termed spectral analysis, permits the Doppler shift frequencies and therefore the blood cell velocities to be displayed as a function of time.

Spectral analysis results in a computer generated gray-scale display of the entire frequency spectrum of the audible Doppler signal. The number of Doppler shifted frequencies present in the signal depends on the velocity distribution of the scatterers in the vessel which, in turn, depend on the cardiac cycle. Since the heart is a pump with a systolic and diastolic component, the gray-scale display will normally exhibit a profile similiar to that shown in Fig. 13-12.

In larger vessels, blood cells near vessel walls may barely move, while those near the center encounter minimal resistance and experience normal laminar flow rates (Fig. 13-13, Region 1). Regions of stenosis in vessels produce high velocity jets of flow that result in large Doppler frequency shifts (Fig.

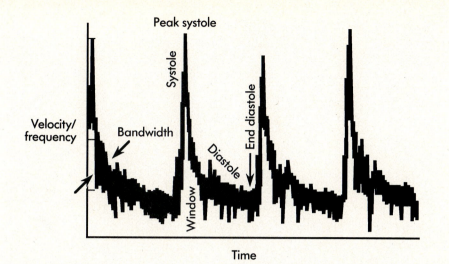

Fig. 13-12 The velocity/frequency spectrum produced with Doppler ultrasound. The display gives a distribution of blood cell velocities in a vessel during the cardiac cycle. Maximum flow rates occur during systole; the minimum during diastole.

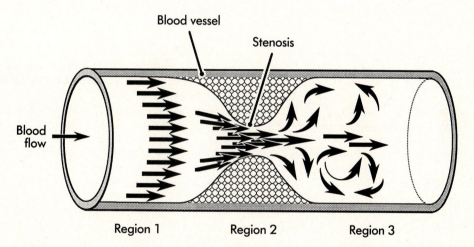

Fig. 13-13 Relation of red blood cell flow during cardiac cycle to the Doppler frequency spectrum. In Region 1, indicative of large vessels, the flow is nearly constant except near the walls. This corresponds to a narrow range of Doppler shifted frequencies. Stenosis (Region 2) causes a dramatic increase in velocity (high peak). In Region 3, whirls and eddies of many different velocities tend to broaden the Doppler spectrum.

13-13, Region 2). Beyond the stenosis, eddy currents are present that include turbulent and reverse flow (Fig. 13-13, Region 3).

As shown, important information about the flow of blood through a vessel can be obtained from observing the bandwidth or spread of frequencies in

the spectral waveform. Normal or laminar flow produces a narrow range of velocities (most of the blood cells move at the same speed) and therefore have a smaller bandwidth. Increased flow rates at a stenosis site produce a wider range of velocities and some spectral broadening. Doppler samples placed im-

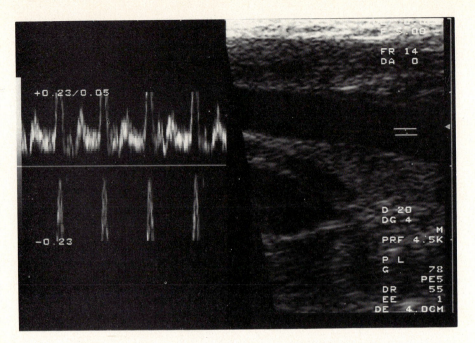

Fig. 13-14 Aliasing has a distinct appearance on the spectral display.

mediately distal to the stenosis will exhibit a wide range of frequencies (marked spectral broadening) and may show some reversed flow.

Aliasing, which occurs when the pulse repetition frequency of the instrument is less than twice the maximum Doppler shifted frequency, produces a distinct artifact on the spectral display. The tops of the velocity/frequency peaks are chopped off and appear to wrap around the bottom of the chart as shown in Fig. 13-14. This occurs because aliasing makes the higher frequencies seem to decrease with increasing flow velocity.

Duplex doppler

The duplex doppler system combines Doppler and real-time B-mode transducers into one probe, so that Doppler flow information can be provided simultaneously with a real-time image. Most duplex systems automatically switch between the pulse-echoB-mode and the pulse-Doppler mode, updating the real-time image several times each second. The image is formed in the conventional way by measuring the amplitude of echoes. Flow rates are determined by measuring the Doppler shift frequency; flow direction is determined by the phase of the returning echoes.

With a duplex system, the operator can isolate the area of interest on the gray-scale image. All velocities within this region can be detected and displayed in time over the cardiac cycle. Regions of stenosis are visualized by examination of the velocity spectrum. Beyond the stenotic area, turbulence and eddy currents produce irregular waveforms with reverse flow information appearing below the zero line.

Color flow devices

Recently developed color-flow Doppler systems provide flow information throughout the gray scale field. These systems employ a dual real-time and Doppler transducer, or array, with sophisticated electronics to detect and process the amplitude, phase, and frequency of returning echoes. In some systems, short pulses are used for the gray-scale image while longer pulses are provided for Doppler sampling. Color Doppler uses multiple sample gates along each line of the image to determine the average flow rate and direction in each part of the real-time image. Returning Doppler shifted frequencies are assigned

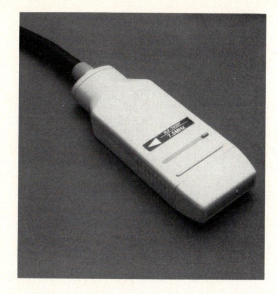

Fig. 13-15 Transducer used with a modern Doppler color flow system. (Toshiba America Medical Systems, Inc.)

a color, either red and blue, as the indicator of flow direction. The intensity of the color indicates the degree of the frequency shift and therefore the magnitude of the velocity of blood. Most manufacturers offer color reverse, since the issue of preferred color versus flow direction has not been completely settled. A typical color-flow transducer probe is shown in Fig. 13-15.

Color-flow devices have several limitations. Only one color can be used at each location to represent the flow at any time. Since the flow distribution in a vessel normally consists of many velocities, the color presented on the display must represent an average of these velocities. Furthermore, the color-flow image usually requires a lower frame rate than a pulse-echo image. The color unit must emit several pulses along each scan line to determine the mean velocity at each point in the line. The sonographer can increase the frame rate by reducing the field of view.

Finally, when a sonographer changes the transducer's transmitting direction, the displayed color representing a particular flow direction in a vessel may also change. The color flow instrument evaluates direction of motion only with respect to the direction of the incident ultrasound beam.

On the positive side, color-flow imaging systems provide important flow information and have been proven very useful in several major applications—the detection and measurement of vascular stenosis, the non-invasive evaluation of organ perfusion, and the visualization of tumor neovascularity. However, as with other Doppler modalities, both knowledge of the fundamental principles of Doppler physics and operator expertise are required to successfully use color-flow Doppler instrumentation.

Review Questions: Chapter 13

1. Define or otherwise identify:
   a. The Doppler effect
   b. Doppler shift frequency
   c. Doppler angle
   d. Quadrature phase detection
   e. Aliasing
   f. Nyquist limit
   g. Spectral analysis
   h. Color-flow imaging

2. Briefly describe how the Doppler effect is used to detect the motion of body structures.

3. Doppler ultrasound results are sometimes presented audibly. Briefly explain how the audible signal is obtained.

4. The magnitude of the Doppler shift depends on the cosine of the angle between the sound beam and the direction of motion. Explain this statement.

5. A 10 MHz transducer scans a soft tissue interface that has a velocity of 10 cm s^{-1} toward the detector.
   a. What is the Doppler shift frequency?
   b. What is the frequency seen by the detector?

6. Answer the questions in the previous problem if the Doppler angle is 45°.

7. Why are two crystals required for continuous wave Doppler? How does one obtain bidirectional CW Doppler information?

8. What is the primary disadvantage of the CW Doppler system? How can this be overcome? What is a potential advantage of CW Doppler?

9. Why is it not possible for a pure Doppler system (CW or pulsed) to provide a gray-scale image of the anatomy? How does duplex Doppler solve the problem?

10. Discuss range gating in pulsed Doppler.

11. At what Doppler angle will the Doppler shift be a maximum? A minimum?

12. Verify (Fig. 13-9) that the range gate for an interface 1 cm from the transducer should be set at 13 μs (roundtrip).

13. How does aliasing appear on a spectral display record? Why?

14 Image Analysis

There was a considerable discussion in Chapter 9 of the three principle characteristics of any medical image—spatial resolution, contrast resolution, and noise. There is a fourth characteristic to be considered during image analysis, artifacts. **An artifact is any unintended information on an image that does not represent the object.** In a sense, an artifact contains negative information because it will complicate and confound the proper interpretation of the image. An artifact is, therefore, a false feature of an image caused by equipment deficiencies, peculiar patient structural features, patient instability, or image processing. Diagnostic sonographs take pulse-echo signals that are time-based and turn them into distance-based images. Therefore any disturbance in the time related echo-signal detection will result in distance related tissue structures that do not represent real anatomy. Such false image structures are sonographic artifacts.

Radiologic technologists are quite familiar with the classical artifacts associated with positioning, radiographic technique, and processing of radiographs. Interpretation of CT and MR images are also complicated by distinctive artifacts. Diagnostic ultrasound is no exception. Proper recognition of the characteristic sonographic artifacts is important for improved image interpretation and reduced diagnostic error.

There are different ways to characterize sonographic artifacts, such as by cause or appearance. Here a combination of the two is used, classifying artifacts by reverberation, shadowing, enhancement, displacement, distortion, and aliasing.

REVERBERATION

When an ultrasound beam encounters two highly reflective interfaces that are close to one another, as illustrated in Fig. 14-1, a portion of the beam will be reflected to the transducer from both the proximal and distal interfaces, providing an accurate image of each. However, some of the beam reflected from the distal interface will be reflected again at the proximal interface and not returned to the transducer until reflected yet a second time from the distal interface. When this twice-reflected beam arrives at the transducer, the echo appears to originate from a depth determined by the total beam travel. This reverberation of the ultrasound between two interfaces can occur many times resulting in an image characterized by repetitive interfaces occurring at regular intervals and decreasing in intensity with depth. These false interfaces are termed a **reverberation artifact,** and a severe example is shown in Fig. 14-2. This type of artifact occurs only when the ultrasound beam is perpendicular to the involved interfaces.

Most reverberation artifacts occur when a highly reflective interface is located near the patient surface, such as a loop of gas filled bowel or superficial abdominal fat layers that are highly reflective. Other more subtle reverberation artifacts can occur between two interfaces located at a depth, but in such cases the appearance of the reverberation artifact is not so prominent. Fluid-soft tissue interfaces such as placenta-amniotic fluid, bladder wall-urine, and gallbladder are examples of deeper structures that can produce the reverberation artifact. Some reverberation artifacts appear as distinct, regularly spaced, repeating echoes, while in others the echo chain is much closer, less distinct, and diffuse. The distinct echoes may be followed by a "ring down" artifact, which is a tail of very fine echoes produced from small gas bubbles.

Occasionally the reverberation artifact will result in a pseudomass that appears either cystic or solid. Pseudomasses can occur when deep abdominal or pelvic gas-filled structures are imaged. The proximal wall of the pseudomass is formed by reflections from the soft tissue-gas interface. The distal wall is a result of reverberation echoes.

A particularly characteristic reverberation artifact called a **comet tail** is sometimes encountered. The process for producing a comet tail artifact is illus-

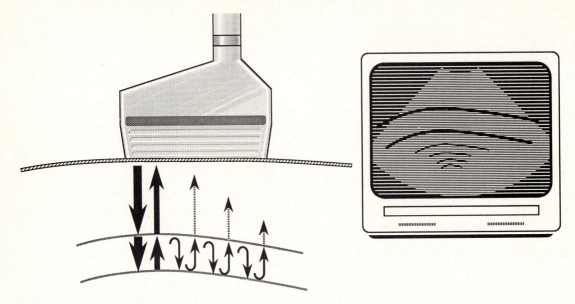

Fig. 14-1 The reverberation artifact results when the ultrasound beam bounces back and forth between two reflective interfaces.

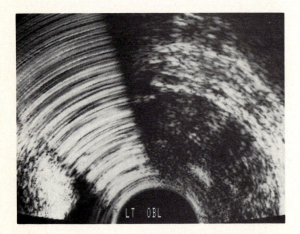

Fig. 14-2 This axial prostate scan shows severe reverberation artifact due to air bubble trapped between transducer and condom.

trated in Fig. 14-3. This artifact generally occurs when two highly reflective interfaces are close together so that multiple reverberations merge and rapidly diminish in intensity on the image. The comet tail is similar to the ring down artifact in appearance, but it is nonrepeating. It is produced when imaging such objects as buckshot, gallstones, gas bubbles, and even surgical clips, as shown in Fig. 14-4.

Reverberation artifacts can sometimes be distinguished from true anatomy by simply repositioning the transducer so that the beam is not perpendicular to the involved interfaces. Reducing the proximal TGC gain and delaying the distal TGC gain through the involved interfaces will also suppress or remove reverberation artifacts. Reverberation artifacts that appear on the image distal to highly reflective interfaces should be investigated further to ensure that they are not confused with real anatomy.

SHADOWING

There are two types of reflections in diagnostic ultrasound—diffuse and specular. **Diffuse reflections** are rather weak in intensity and fuzzy in appearance. They occur because of multiple scattering of the ultrasound beam, including backscattering rather than reflection in the normal sense. The tissue in-

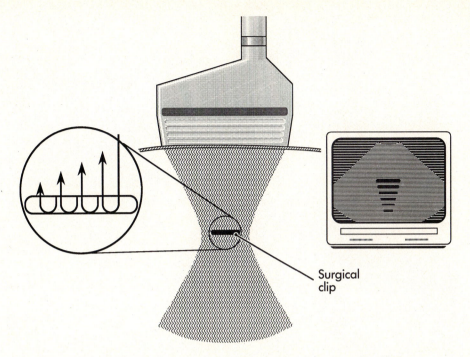

Fig. 14-3 Schematic drawing of the production of a comet tail artifact.

terface producing the diffuse reflection will be irregular and rough. As the degree of roughness increases the magnitude of the scattering process increases, and the intensity of the diffuse reflection
will be less.

Reflective shadows

Specular reflections are those occurring from a
large, smooth tissue interface separating tissues having considerable difference in acoustic impedance.
They are the most efficient type reflections and can
be responsible for a **shadowing artifact.**

When the ultrasound beam is incident on a highly
specular, reflective interface such as soft tissue (gas)
shown in Fig. 14-5, nearly all of the beam intensity
will be reflected. Very little will be transmitted across
the interface so that essentially no signal is available
to interact with deeper structures. The effect is to
remove real structures from the image. Acoustic
shadows produced by specular reflective interfaces
are often not complete shadows, but have some signal contained in them such as weak reverberation
and ring down artifacts.

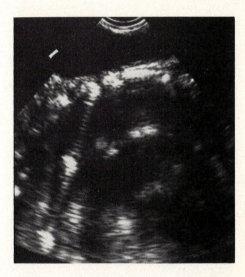

Fig. 14-4 This comet tail artifact in a longitudinal pelvis
image results from ring down from a surgical clip.

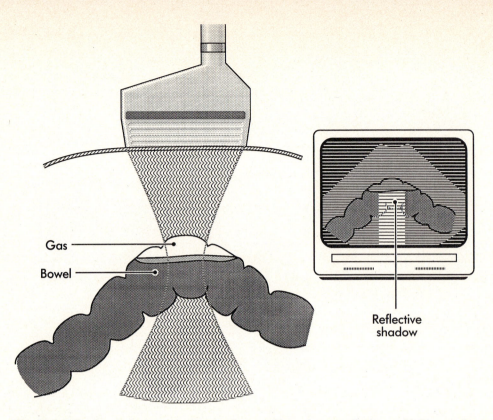

Fig. 14-5 Reflective shadows occur when the ultrasound beam is perpendicular to a highly specular reflecting interface such as a gas-filled loop of bowel.

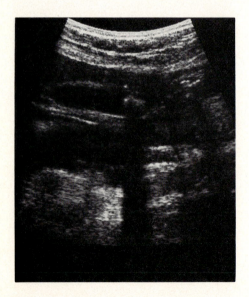

Fig. 14-6 The gallstone in this sonograph created a thin reflective shadow artifact.

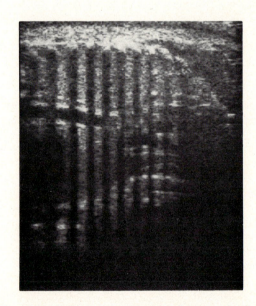

Fig. 14-7 Rib shadows in this coronal image of a fetus produced this multiple shadow artifact.

Attenuation shadows

Shadowing can also result from highly attenuating tissue such as compact bone. Bone, the most rigid tissue in the body, has attenuation properties twenty times as high as soft tissue. Because of the high attenuation of ultrasound in bone, reverberation artifacts are absent in the shadow. Gallstones and other calcified masses can also produce acoustic shadows by attenuation as seen in Fig. 14-6. The artifact is not so distinct, but it does obscure visualization of real structures. This type of shadowing artifact can sometimes be removed by increasing the late TGC. Fig. 14-7 is another example of such a shadowing artifact, which obviously obscures examination of the fetal chest. These are rib shadows in a coronal image of a fetus.

Edge shadows

An additional type of reflective shadow can occur if the soft tissue-gas interface is curved sufficiently. The incident beam is reflected in such a manner that none of it returns to the transducer. The result is an **edge shadow.**

If the rounded structure that intercepts the ultrasound beam is not reflective, but rather refractive, then a refractive-edge shadow can be produced. Refractive-edge shadows occur when the velocity of sound in the structure differs significantly from the velocity of sound in surrounding tissue. Snells Law becomes operative, causing refraction at the edge of the structure.

Fig. 14-8 shows what happens when the velocity of sound in surrounding tissue is higher than that in the structure. Cystic lesions in liver are such an example. In a fluid filled cyst ultrasound velocity is approximately 1500 m s^{-1}, while in liver it is approximately 1550 m s^{-1}. The rays of the ultrasound beam are compressed, and a narrow shadow occurs.

If the velocity of sound is higher in the structure than in the surrounding tissue, as seen in Fig. 14-9, then the opposite occurs. The rays of the ultrasound beam are spread out producing a wide shadow. Ultrasound has velocity of approximately 1500 m s^{-1} in amniotic fluid. When an ultrasound beam encounters fetal skull, where its velocity is approximately 3000 m s^{-1}, the ultrasound beam is spread out producing a wide shadow.

Fig. 14-10 is yet another example of an edge shadow. Here the ultrasound beam is incident at the **critical angle** of a gall bladder and edge shadowing results. Another critical angle artifact is shown in

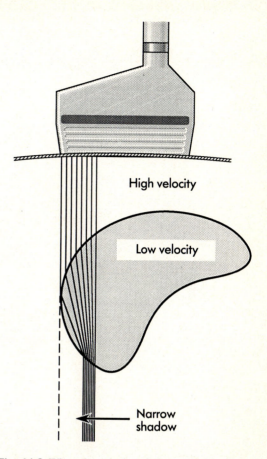

Fig. 14-8 When the velocity of ultrasound in a structure is lower than that in the surrounding tissue, a narrow refractive shadow occurs.

Fig. 14-11. This artifact, seen on an image of the left portal vein, has been observed only in Texas!

ENHANCEMENT

Enhancement occurs because the ultrasound imager anticipates there will be attenuation related to depth of penetration in soft tissue. In order to maintain uniform brightness, the imager alters the presentation of echoes from deep structures, so that there is more uniform image intensity. When echoes pass through cystic structures, as in Figs. 14-12 and 14-13, this attenuation does not occur. There is a false amplification, which results in the appearance of an enhanced signal. The imager anticipates a loss of echo amplitude, but it does not happen. Therefore,

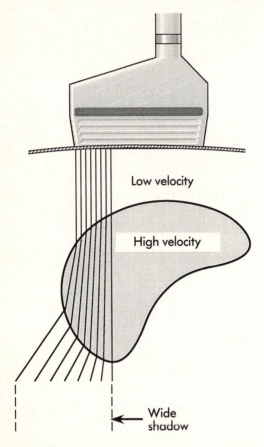

Fig. 14-9 When the velocity of ultrasound in a structure is higher than that in the surrounding tissue, a wide refractive shadow occurs.

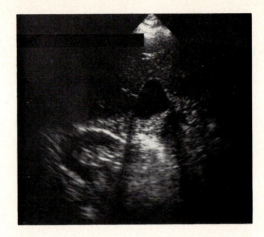

Fig. 14-10 Edge shadows produced at the critical angle of a gallbladder.

there is false amplification of distal tissue resulting in enhancement.

DISPLACEMENT

Constructing an ultrasound image requires geometric simplicity between the ultrasound beam and the patient. The beam should be straight, not refracted, and the reflection should be directed straight back to the transducer, not along a devious path.

There are three imaging characteristics that can result in a spatial shift—**misregistration, multipath reflections, and beam-width distortions.** A spatial shift occurs when structures in the beam are moved to a displayed position outside of the beam, or when structures that are actually outside of the

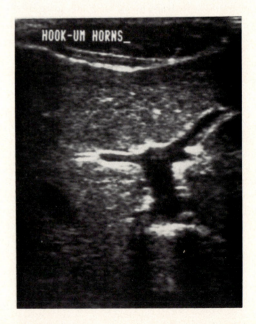

Fig. 14-11 This left portal vein artifact edge shadow is seen only in Texas.

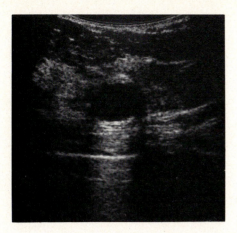

Fig. 14-12 Enhancement of tissue distal to a renal cyst.

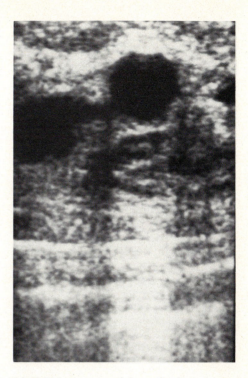

Fig. 14-13 Enhancement of tissue distal to a breast cyst.

beam are made to appear inside the beam. For articulated arm B scanners, misregistration is always a concern. The orientation of the ultrasound beam in some structures will cause multipath reflections returning to the transducer. The ultrasound beam width should be narrow and contain no side lobes in order to minimize artifacts.

Misregistration artifact

During static B-mode imaging with an articulated arm, there should be proper registration between the position of the ultrasound beam and its indicated position on the final image. Accurate registration requires calibration of the electrical potentiometers in the arm, proper electronic signal processing, and a precise time-base configuration. Simple routine quality-assurance exercises, such as those described in Chapter 15, should be instituted to prevent misregistration artifacts. Such artifacts can appear as a juxtaposition of tissues, but they usually appear as a general degradation of image quality by blurring. The transducer scanning motion can influence the degree of the misregistration artifact.

The importance of misregistration artifacts is insignificant at the present time, and this discussion is given principally for historical interest. **Misregistration artifacts do not occur in real-time imaging.**

Multipath artifacts

The sonographer manipulates the transducer so that the ultrasound beam will interact with tissue interfaces and reflect directly back to the transducer. However, this can only occur if the reflecting interface is perpendicular to the path of the transmitted ultrasound beam. If the interface is at an angle, the reflected beam can be directed away from the transducer and escape detection altogether. This process is illustrated in Fig. 14-14. In such a situation no multipath artifact is produced.

However, if there are multiple surfaces or a continuous curved concave surface, such as that shown in Fig. 14-15, the ultrasound beam can undergo multiple reflections and still return to the transducer. This multiple reflective path takes longer than a single reflection and can therefore produce false structures at greater depths. This is a **multipath artifact.** Fig.

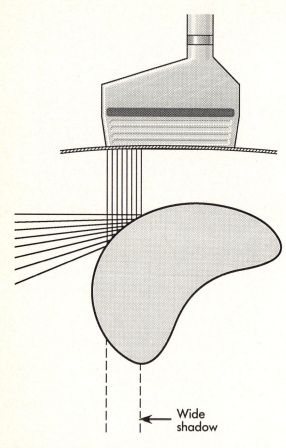

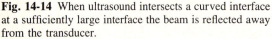

Fig. 14-14 When ultrasound intersects a curved interface at a sufficiently large interface the beam is reflected away from the transducer.

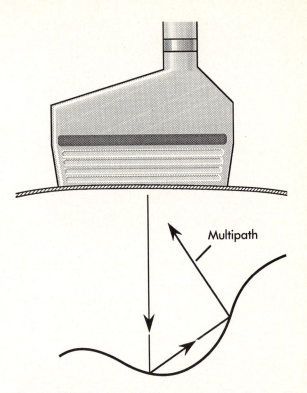

Fig. 14-15 Multipath artifacts occur when the ultrasound beam undergoes two or more reflections and then returns to the transducer.

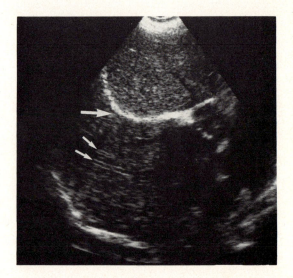

Fig. 14-16 A multipath artifact in the form of false deep structures can occur as in this longitudinal image of the liver demonstrating false portal veins (*small arrows*) on the wrong side of the diaphragm (*large arrow*).

14-16 is a longitudinal image of the liver demonstrating false portal veins as artifacts.

Any concave specular reflecting surface, such as the bladder, can produce multipath artifacts. Fortunately, the artifact can be reduced or removed by reorienting the transducer.

Beam-width artifacts

Ideally one would like to have an extremely thin ultrasound beam that would result in better lateral resolution and improved image quality. However, ultrasound beams are not only thick, but the beam width varies with depth, especially for focused trans-

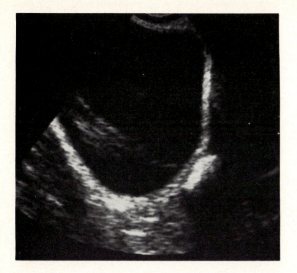

Fig. 14-17 The variable beam width with depth of most transducers can result in a beam width artifact such as sludge in the gallbladder.

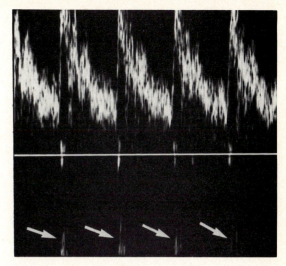

Fig. 14-18 Pulsed doppler spectrum of common carotid artery artery that demonstrates aliasing at peak velocities *(arrows)*.

ducers. This characteristic can result in the appearance of sludge or other particulate material in the gallbladder as seen in Fig. 14-17, when in fact none exists. If the focal length of the transducer is at the middle of the gallbladder, the width of the beam at the posterior surface will be wider. The result is a partial volume effect along the posterior wall of the gallbladder giving the appearance of particulate matter in the gallbladder. Images of other fluid-filled, specular reflecting tissues, like the bladder and cysts, can also contain this beam-width artifact.

When scanning obliquely through the body, the change in beam width can cause some structures to appear displaced from their normal location. This displacement artifact can be accentuated by side lobes. All ultrasound beams have side lobes as described in Chapter 8. They result from diffraction of the transmitted ultrasound beam as it passes through tissue. The side lobes rarely have intensity greater than 1%, −20 dB, of the central axis beam and therefore are normally of no consequence. Under some circumstances, however, they can accentuate the beam width, destroying lateral resolution and contributing to the displacement artifact.

The side lobe intensity is transmitted at a different angle from the central axis. When it intercepts the interface in question, the echo will return early or late depending upon the curvature of the surface. The result is a false image of the surface, appearing either in front of or behind the true interface. This beam-width artifact can sometimes be controlled by transducer positioning or by reducing the transducer's output power.

DISTORTION

Distortion in a sonograph is considered an artifact because the object is falsely indicated because of beam characteristics or electronic insufficiency. The artifact produced is either **geometric distortion** or **signal intensity distortion.**

The variation in ultrasound beam width and depth into the patient can create geometric distortion. Objects outside of the plane of focus appear larger than normal because of the thicker beam width. Consider a situation that occurs when imaging a point. A narrow beam will produce a more accurate image of the point than a broad beam. This occurs because the lateral resolution cannot exceed the width of the ultrasound beam. This type of artifact is usually controlled by proper motion of the transducer.

An additional type of beam-intensity artifact can occur with systems using a digital scan converter with too few gray levels. Because of the limited gray scale, adjacent pixels are assigned the same intensity when they should be slightly different. The result is a contouring effect that can simulate structures within the image.

ALIASING

Real-time imaging is helpful because of the ability to view moving structures. Fetal imaging near the end of term will generally show a lot of motion, especially the fetal heart. If the real-time frame rate is too low, the image will not accurately represent the true motion because of aliasing.

Aliasing is an artifact that deals with the rate at which a moving object is sampled. In general, if one does not sample at a frequency at least twice that of the object's motion, an aliasing artifact will occur. Fig. 14-18 is a Doppler frequency spectrum of arterial blood flow. Peak velocities are so high they are aliased into the negative portion of the frequency spectrum.

Computed tomography and magnetic resonance imaging can suffer greatly from aliasing artifacts. Aliasing is not as big a problem in ultrasonography.

Doppler imaging, especially pulsed Doppler, is the main ultrasound examination that can be subject to aliasing artifacts. Where pulse-echo imaging involves simply the time domain, Doppler adds frequency and amplitude of frequency, which compound aliasing artifacts.

Review Questions: Chapter 14

1. Define or otherwise identify:
   a. Artifact
   b. Reverberation
   c. Pseudomass
   d. Specular reflection
   e. Misregistration
   f. Aliasing

2. Name four of the six classifications of sonographic artifacts.

3. Which type of artifact results in a comet-tail appearance?

4. What are the three types of shadowing artifacts and which is most bothersome?

5. List the three types of displacement artifacts.

6. The aliasing artifact is most often observed in what type of examination?

15 Quality Control

Every field of medicine today is required to develop and conduct programs to ensure the quality of patient care and management. Diagnostic imaging is a leader in promoting such programs for all of the subfields, including diagnostic ultrasound. Two of these programs are **quality assurance (QA)** and **quality control (QC).** There is a subtle difference between the two. **Quality assurance deals with people,** and its programs evaluate the scheduling of the patient, the accuracy of image interpretation, the dispatch of the diagnostic report, and the overall impact of the procedure on patient response. **Quality control programs deal with instrumentation and equipment,** the subject of this chapter.

In addition to care of the patient, there are other reasons for conducting a quality control program in diagnostic ultrasound. Our litigious society demands records pertaining to quality control. Some insurance carriers will pay for services only from facilities performing quality control. The Joint Commission on Accreditation of Healthcare Organizations (JCAHO) will not place its seal of approval on facilities that do not have an on-going quality control program.

In diagnostic ultrasonography a quality control program contains three principle areas: assessment of diagnostic accuracy, maintenance of imaging apparatus, and periodic equipment-performance monitoring. We deal with only the latter, **periodic equipment-performance monitoring.** Such a monitoring program should be conducted with the same test objects to ensure the consistent fidelity of the resulting image. Under some circumstances, one may wish to monitor the physical characteristics of the ultrasound beam, such as spatial pulse length and output intensity, but normally such measurements are left to the manufacturer. Equipment-performance monitoring is based on image analysis.

The equipment to be periodically monitored includes not only the ultrasound console but also the various ultrasound transducers, the hardcopy imager, and the image processor. Some measurements would

be performed only during acceptance testing of new equipment, but most should follow a daily, weekly, or monthly schedule.

To monitor performance of diagnostic ultrasound systems, phantoms or test objects are used, but there is a distinct difference between the two. **Phantoms are anatomy-simulating devices, while test objects are usually geometric configuration** designed to evaluate specific system parameters. In radiography and radiation oncology, phantoms find considerable use. Quality control in diagnostic ultrasound relies principally on test objects.

QUALITY CONTROL TEST DEVICES

Over the years a number of test objects have been designed to measure various QC performance parameters. The box on page 144 lists some of the more popular test devices currently available. Each has characteristics providing certain specific advantages, however rarely does a diagnostic ultrasound-imaging department require more than one or two such test devices. The AIUM test object is perhaps the most popular and therefore the one discussed here in detail. Others will be discussed but principally for identification. The manufacturers identified in the box on page 144 will provide complete information on any devices requested.

AIUM test object

The AIUM test object was designed in 1975 by a task group of the American Institute of Ultrasound in Medicine and has since served as the standard quality control-test object. Although both an open and closed version of this test object are available, the closed version is more popular. It consists of a five-sided plastic block with outside dimensions of $160 \times 140 \times 50$ mm, with one beveled corner. Stainless steel or plastic 0.75 mm diameter reflecting rods are arranged in a 100×100 mm grid. The open version of this test object is usually filled with an alcohol-water mixture laced with an algae inhibitor so that the velocity of ultrasound is 1540 m s^{-1},

AVAILABLE TEST DEVICES FOR USE IN A DIAGNOSTIC ULTRASOUND QUALITY CONTROL PROGRAM (SUPPLIER*)

Test objects

American Institute of Ultrasound in Medicine (AIUM) Test Object (RMI)
Dynamic Doppler Flow System (NA)
Beam Shape Phantom (NA)
Sensitivity, Uniformity, and Axial Resolution Phantom (NA, RMI)
Scan Plane Thickness Phantom (RMI)

Phantoms

Multi-purpose Tissue/Cyst Phantom (NA)
Tissue Equivalent Phantom (NA)
Contrast/Detail Phantom (NA)
Prostate Phantom (NA)
Particle Image Resolution Test Object (NA)
Tissue Mimicking Phantom, five models (RMI)
Doppler Phantom/Flow Control System (RMI)

---

*NA = Nuclear Associates
RMI = Radiation Measurements Inc.

the same as that in soft tissue. The closed version is also constructed so that the velocity of sound is 1540 m s^{-1}. The AIUM test object is shown schematically in Fig. 15-1.

The array of rods consists of five sets identified A through E. Each group of rods has a specific purpose. For ease of identification, the AIUM test object is viewed with the truncated corner to the upper left.

Group A: This group consists of a vertical row of six rods spaced 20 mm apart, and is used to evaluate depth calibration, vertical linearity, and gain.

Group B: The six rods along the bottom of the phantom are similar to group A in that they are spaced 20 mm apart. These rods are used for evaluating sensitivity, horizontal calibration, and horizontal linearity. When imaged through several surfaces groups A and B can also be used to evaluate registration of a static B-scanner.

Group C: This is a group of seven rods spaced vertically at 25, 20, 15, 10, 5, and 3 mm increments. These rods are scanned from the side and are used to evaluate lateral resolution and beam width.

Group D: There are four rods in group D positioned at 2, 4, 6, and 8 mm depths respectively, from the top surface of the test object. This group is designed to evaluate the dead zone of a transducer.

Group E: Group E rods are those positioned in the center of the test object. The top rod is at the geometric center of the 100 mm by 100 mm array of test groups, and there are five additional rods lying in a plane tilted 15° from the center line at 5, 4, 3, 2, and 1 mm spacings. Placing the rods at an angle reduces interference from shadowing and reverberation artifacts. These rods are designed to measure axial resolution and registration.

The routine use of the AIUM test object is quite simple. A number of measurements can be made and the results recorded in just a few minutes. This test object does not deteriorate with use and can withstand rather vigorous handling. The box on page 145 summarizes the ten elements of a QA program that should be routinely performed with the AIUM test object.

SUAR test object

The **s**ensitivity, **u**niformity, and **a**xial **r**esolution (SUAR) test object is a device designed to measure these three diagnostic ultrasound properties. It is also an acrylic block measuring 160 × 140 × 45 mm with a 100 mm long wedge cut out of its bottom. This test object is shown schematically in Fig. 15-2. The wedge compartment is filled with a fluid having a sound velocity of 1540 m s^{-1} for precisely evaluating sensitivity and axial resolution.

A thermometer is incorporated to monitor the temperature of the block. The temperature must be within the range of 20° to 22°C in order for the velocity and attenuation of ultrasound to remain constant. As temperature increases, the attenuation of ultrasound by the acrylic block increases. If the temperature variation exceeds 5°C from one evaluation

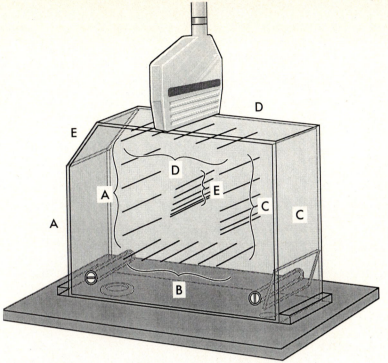

Fig. 15-1 This schematic rendering of the AIUM test object identifies the five separate groups of rods and the four surfaces for conducting quality control measurements.

QUALITY ASSURANCE TESTS THAT SHOULD BE ROUTINELY PERFORMED WITH THE AIUM TEST OBJECT

1. System sensitivity
2. Distance or time calibration
3. Accuracy of distance caliper
4. Evaluation of misregistration
5. Determination of dead zone depth
6. Evaluation of axial resolution
7. Evaluation of lateral resolution
8. Determination of time-gain characteristics
9. Evaluation of display system—CRT, hardcopy camera, processor
10. Visual inspection of all components

to the next, a temperature correction factor will be required.

Sensitivity. The 140 mm and 160 mm dimension of the block are used to evaluate the overall sensitivity of the imager. When the transducer is acoust-ically coupled to either of these faces, an image of an acrylic/air interface is produced. Usually lower frequency transducers will be evaluated along the 160 mm dimension and higher frequency transducers along the 140 mm dimension. This represents a total transit distance of 32 cm and 28 cm respectively, very representative of the maximum distance encountered clinically.

Choose the path through the block that will easily image the interface at the normal setting of the sensitivity controls. These controls may include output control, time gain compensation, and amplifier gain. Adjust one or all of these sensitivity controls so that the image of the acrylic/air interface is just barely visualized. Record the values of each control. **Daily variation should not exceed 3 dB.**

Uniformity. The flat, machined surface of this acrylic test object should provide a uniform signal intensity when imaged. The image should appear as a uniform line of brightness for both B-mode and real-time transducers. The test is more helpful for real-time transducers because use of static B-mode requires a very steady motion of the transducer.

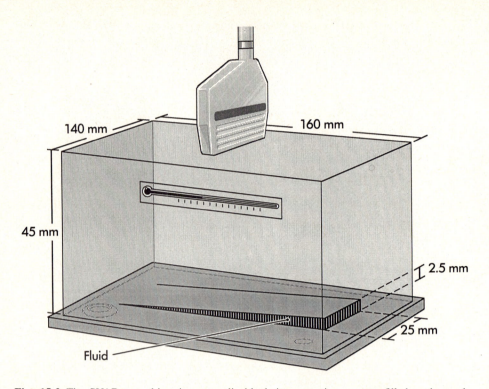

Fig. 15-2 The SUAR test object is an acrylic block incorporating a water-filled wedge and a thermometer.

To perform this test the transducer must be well coupled to the acrylic block with gel. Any air bubbles will interfere with the evaluation. The system sensitivity controls should be adjusted for echoes that are barely perceptible, and the control positions recorded. For a sector scanner the lateral echoes may not be imaged even though the central axis echoes are imaged. In such a situation a gain control will have to be increased in order to view these lateral echoes. The difference in gain setting between the central axis and the lateral echoes represents the degree of nonuniformity. **Nonuniformity should not exceed 15 dB because this level will be clinically noticeable.**

Axial resolution. The water-filled wedge at the bottom of the test object, a depth of 4 cm, is used to evaluate axial resolution. The transducer is carefully coupled to the top surface of the test object and scanned in such a fashion that both surfaces of the wedge are imaged.

Each wedge surface is an acrylic/fluid interface

that is similar in reflectivity to that encountered clinically. The wedge is 100 mm long, and the two interfaces rise from a 0 mm to a 2.5 mm separation. The rate of rise is therefore 0.25 mm per 10 mm along the length of the wedge. When imaged, axial resolution is evaluated by the minimum separation of the two wedge surfaces that can barely be resolved. This value is determined by measuring the lateral distance along the wedge, and multiplying this distance by the rate of rise to obtain the axial resolution. The result should be recorded.

Example: The length of an SUAR wedge image that appears as two interfaces is 43 mm. What is the axial resolution?

Answer: 43 mm $\times$ .25 mm/10 mm = 1.08 mm

RMI tissue phantoms

Radiation Measurements Incorporated (RMI) has several popular phantom models identified as tissue-mimicking phantoms. These phantoms are designed

Fig. 15-3 Four popular tissue mimicking phantoms. (Courtesy RMI, Inc.)

with sets of discrete line targets and simulated cysts embedded in material having sound velocity, attenuation, and scattering properties similar to soft tissue. Fig. 15-3 is a photograph of four of these phantoms and a flow test object.

One advantage of these phantoms is that quality control measurements are made at scanner settings used for clinical imaging. The phantoms differ in size and number of test objects embedded in the tissue mimicking material. Hard copy images are retained to check constancy of gray-scale photography, depth of penetration, registration accuracy, resolution, and cyst imaging.

Fig. 15-4 is a schematic of one of these tissue mimicking phantoms that illustrates the complexity of the phantom. The material itself is a water-based gel with microscopic graphite particles mixed throughout. The ultrasonic velocity in the gel is controlled to 1540 m s^{-1}, and the ultrasonic attenuation is adjusted to 0.5 to 0.7 dB cm^{-1}MHz^{-1}. Because this gel also has the same frequency dependence as soft tissue, this phantom can be used with essentially all transducers.

Throughout the gel are small particles of scattering material causing an echogenic image similar to soft tissue. Simulated cysts consisting of low attenuating, scatter-free cylinders are positioned in the gel and should appear echo free. In addition, as in the AIUM

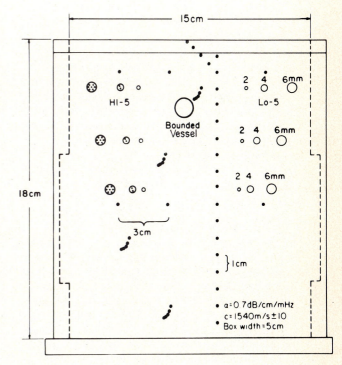

Fig. 15-4 Schematic drawing of RMI model tissue mimicking phantom.

Table 15-1 Use of the AIUM test object for a ten-point quality control program

Performance evaluation	Image rod group	Through surface
Sensitivity	B	D
Distance calibration	A,B	A,C,D
Caliper calibration	A,B	A,C,D
Registration	E	A/C and D/E
Dead zone	D	D
Axial resolution	E	D
Lateral resolution	C	A,C
Time-gain compensation	A	D
Display characteristics	E	C,D
Visual inspection		

test object, rod reflectors are positioned geometrically.

Although not absolutely necessary, phantoms do find application in quality-control programs for ultrasound. Phantoms are useful for subjective evaluation of ultrasound scattering, attenuation, and specular reflective properties. Some phantoms contain echo-free cystic simulating cavities, echo-free cylinders of several sizes at several depths, scattering layers to demonstrate beam profile characteristics, and various shapes of several different scattering materials to demonstrate the dynamic range of gray scale of the imager.

Such phantom images should be obtained monthly and retained for comparison with earlier images.

ROUTINE MEASUREMENTS AND OBSERVATIONS

There are many performance-measuring observations that can be made on a diagnostic ultrasound unit. Those that follow would adequately constitute a quality control program in diagnostic ultrasound, and all can be done with the AIUM test object. Before beginning any tests be sure that the test object is securely positioned and properly aligned. It is not acceptable to do these tests on soft patient pads or wheeled stretchers. Because of the many probes available, it may not be possible to evaluate each transducer in routine use during each QC test. In such situations the most frequently used probe

should be evaluated each time and other probes less often. Table 15-1 summarizes the rod groups to be imaged and the surface on which to place the transducer for each of the QC tests that follow.

Sensitivity

The sensitivity of an imaging device is evaluated by determining how low a signal or how weak a reflection is detectable. Normally, one would evaluate sensitivity first and then increase the gain settings for subsequent evaluations. Sensitivity can be evaluated by imaging any rod in the block. Usually it is evaluated by placing the transducer on surface D, the top surface, and imaging the lower right rod of group B.

Under normal circumstances one should be able to image this rod easily. Because there is relatively little ultrasound attenuation in the gel of this test object, sensitivity and gain controls will be set to a low level. This rod lies at a depth of just over 100 mm from the transducer face, and therefore the echoes are from a rather deep structure. The time gain compensation control (TGC) should be adjusted so that the rod in question is just barely imaged. Note the position of the TGC.

With time and use, the sensitivity of the system may deteriorate. This is monitored by noting the TGC position during subsequent sensitivity evaluations. As the system ages, more amplifier gain may be required. **A recommended maximum permissible change is ± 3 dB.** This represents about 5 mm loss in depth sensitivity. When acceptance testing a new system, this measurement represents maximum sensitivity.

For the following additional QC measurements, an additional 10 dB or so should be added to the system sensitivity.

Distance calibration

Rod groups A and B are used for this evaluation. Each of these rods is separated by 20 mm. Therefore a vertical image through surface D or a horizontal image through surfaces A or C should result in an array of images also separated by 20 mm. When these rod groups are scanned, a hardcopy image is made, and the distance between rod images is measured to ensure that they are all the same and all have a 20 mm separation. The important characteristics of such a test are precise rod spacing and the speed of sound in the medium. **The total error permitted of the separation of the first and last rods, a distance of 100 mm, is ± 2% or ± 2 mm.**

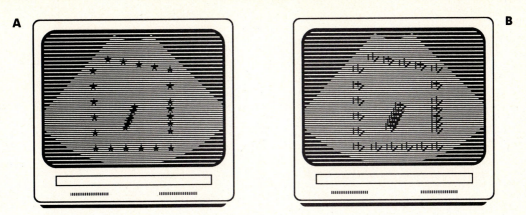

Fig. 15-5 Good static B scan registration will show line images of rods that are perfectly bisected. **A,** good registration. **B,** misregistration.

If the spacing between rods is not uniform, or the spacing of the outside rods is not within the range of 98 to 102 mm, then service is needed.

The following QC tests require that the test object be made of material having ultrasound velocity of $1540 \pm 3 \text{ m s}^{-1}$. Solid test objects are manufactured to this tolerance. A solution-filled test object must be fabricated precisely, but the proper solutions are commercially available.

Caliper calibration

As with distance calibration, caliper calibration employs rod groups A and B. In addition to imaging these groups as before, electronic calipers, if they are available on the system, are superimposed on the image adjacent to the rod groups. A visual analysis can then be made between the separation of the rods and the separation of caliper dots. **Again a ± 2% difference over the 100 mm grid is the maximum allowed.**

Electronic calipers are always on contemporary equipment, but on some older models this evaluation will require hand-held calipers. Should this test fail for a solution-filled test object, it is then necessary to determine if the depth-marker circuitry or the test object is at fault. Measurements should always be made from the center of the leading edge of each rod image, not from the center of the rod image.

Registration

The proper positioning of objects on the image, regardless of the surface through which the objects are imaged, is called **registration.** When objects scanned through different surfaces are not superimposed, they are said to be **misregistered.** This evaluation applies principally to static B scanners and is designed to determine that the probe alignment and position are properly calibrated. Before conducting this test, be sure that sufficient coupling gel is applied to all surfaces of the test object.

Any set of rods can be imaged, but group E is normally employed for this measurement. The B scan probe is scanned over the two vertical surfaces, A and C, and the two top surfaces, D and E. Images are stored from each of these four surfaces and superimposed on one another.

Because of the inherently poor lateral resolution, each rod will appear as a short line. When superimposed, however, the lines should bisect one another. The distance between the centers of the leading edge of each rod image should not exceed 3 mm. Fig. 15-5 shows the difference between (A) good registration, and (B) misregistration.

With a real-time probe, a similar analysis can be conducted to ensure that lateral distances are properly registered. Image rod group B through the center of surface D. Freeze the image and display a line of marker dots spaced at 10 mm intervals next to the rod images. The separation of the most lateral rod images from the marker dots should not exceed 3 mm.

Dead zone

The evaluation of dead zone requires the use of the top row of rods, group D, imaged through surface D. Because of reverberations and the very strong

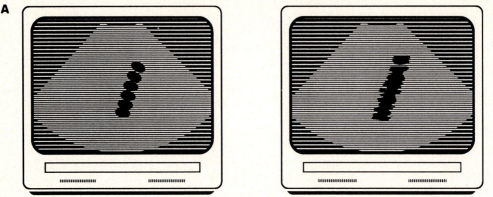

Fig. 15-6 Axial resolution is evaluated by imaging rod group E through surface D. **A,** Good resolution. **B,** Poor resolution.

echo between the transducer and the entrance surface of the test object, a very short region into the test object will produce no signal. This is known as the **dead zone.** Scanning along surface D allows one to evaluate the depth of the dead zone, by noting which of the rods in group D are visible on the image. **The depth of the most shallow rod is the dead zone, and this should not exceed 4 mm for most probes.**

Axial resolution

Axial resolution in diagnostic ultrasound is always better than lateral resolution. The axial resolution is evaluated by imaging rod group E through surface D. Not all rods of group E will be visualized. Axial resolution is considered to be the spacing of the group E rods that are just barely distinguished from one another. Evaluation of axial resolution is a measure of signal processing electronics, transducer crystal integrity, and the various frequency-dependent matching layers. This observation will prove to be a good constancy check also.

The two bottom rods, spaced 1 mm apart, should be clearly visualized. This results in 1.0 mm axial resolution that should be observed with all imagers. Fig. 15-6 schematically shows the difference between good (A) and poor (B) resolution.

Lateral resolution

Lateral resolution is evaluated by scanning the group C rods through both the A and C surfaces. The image produced through surface A is of rods 150 mm deep, that through surface C is of rods 50 mm deep. There-

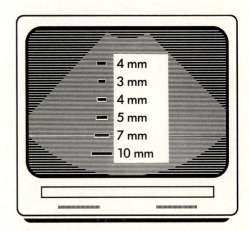

Fig. 15-7 An image of rods in row A through surface D will provide an estimate of lateral resolution as a function of depth.

fore these images provide a measure of lateral resolution at 50 mm and 150 mm. If a focused transducer is used whose focal plane is other than 50 mm or 150 mm, an underestimate of lateral resolution will be obtained. Nevertheless, it should provide a good constancy check.

Such scans will result in an image showing of the rods displayed as a line so that not all the rods may be individually imaged. Lateral resolution is recorded as that between the closest two rods that are separately identifiable. The lower two rods are separated by 3 mm, which is the best lateral resolution measureable with this test object. **All systems should be able to image this 3 mm separation.**

Another measure of lateral resolution can be made using an image of the rods in group A. Scanning this group through surface D will result in lines at various depths separated by 20 mm. A focused transducer will produce an image such as that in Fig. 15-7. Here the depth of focus is shown to be approximately 40 mm where the lateral resolution is measured to be 3 mm.

Because the lateral resolution of a transducer array is dependent upon slice thickness, which varies with depth, test objects have been designed to measure slice thickness. The test object in Fig. 15-8 is a rectangular box containing a diffuse scattering plane at a 45° to the vertical. The top section of the test object contains tissue-mimicking material and the bottom a more dense material.

As one scans the test object shown in Fig. 15-9, the two dimensional image of the ramp changes with depth. The sharpest image, representing the narrowest slice thickness, will occur at the focal distance when the transducer is perpendicular to the long dimension of the top surface. If the transducer is positioned parallel to the long dimension of the top surface, as in Fig. 15-10, a rendering of beam divergence with depth will be observed. This type of test object is particularly helpful in evaluating multifocus and variable focus transducers.

Time-gain compensation (TGC)

The time-gain or distance-gain compensation of an ultrasound imager are evaluated using multiple images of rod group A through surface D. First the rods of group A are imaged with the TGC switch in the off position. The attenuator settings required to just barely display each rod are recorded. This sequence is repeated with the TGC set at a standard level. The difference between the attenuator settings required to display each rod represents the time-gain compensation at each depth. The TGC will vary with transducer and operating frequency, so it should be determined for both.

Display characteristics

Often what is visualized on the video monitor of the ultrasound imager will not be properly recorded on hardcopy. This test is rather subjective and can be done without the test object. One can use a daily clinical image of the liver, a tissue phantom, or a

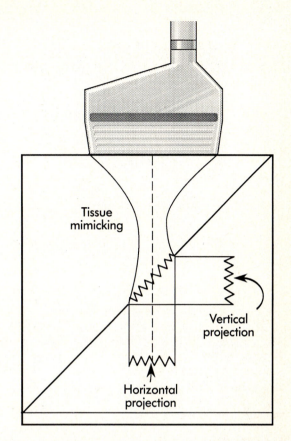

Fig. 15-8 This special test object is designed to evaluate slice thickness for a real-time transducer.

scanner generated gray-bar pattern. When the test object is used, E rods should be imaged through C and D surfaces.

Regardless of which method is used, the hardcopy should then be visually compared with the image on the CRT monitor. Particular attention should be paid to the texture of the gray scale resulting from deep echoes or from small specular reflections. Significant differences need to be investigated to determine if the gray-scale capacity of the camera or the film processing is at fault. With proper QC monitoring of the film processor, one may conclude that any observable differences are attributable to the camera.

Visual inspection

Before each performance evaluation, the operator should make a visual inspection of the scanner, par-

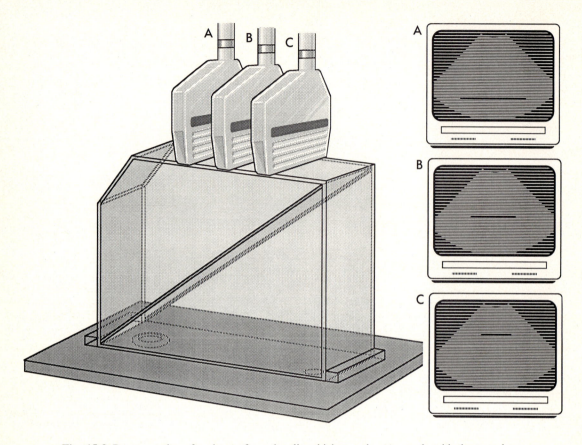

Fig. 15-9 Representation of an image from the slice thickness phantom made with the transducer array perpendicular to the long dimension of the top surface. The narrowest slice thickness occurs at the focal distance.

ticularly the electric safety checks. A volt-ohm meter can be used to verify that the scanner is properly grounded so that during scanning no externally-grounded components touch the sonographer or the patient.

A careful visual inspection of each transducer probe should also be made. Rough handling can cause problems that may first become apparent on visual inspection.

PROGRAM FREQUENCY

For medical-imaging devices, a schedule of the frequency of evaluation (ranging from weekly to monthly) should be developed for various tests. Use of the AIUM test object is simple and requires no

more than 5 minutes for the experienced operator. Most tests should be conducted monthly. A weekly evaluation of distance calibration, axial resolution, and display characteristics is recommended. Table 15-2 is a review of the measurements involved in the ten-point quality-control program, the recommended frequency for each evaluation, and a recommended permissible tolerance.

With the use of the AIUM test object, it is not necessary to make hardcopy images for retention. However, it is not sufficient to say the quality control tests are conducted routinely. A simple form should be used to maintain a running log of the results of this quality-control activity. For completeness, each series of tests must be initialed by the sonographer.

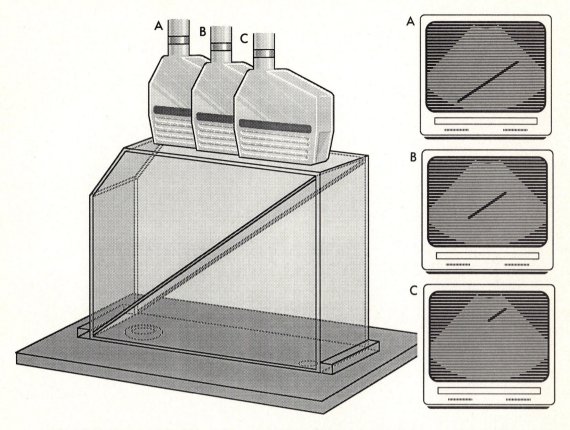

Fig. 15-10 When the slice thickness test object is imaged with the transducer array oriented parallel to the long dimension of the top, the image represents the divergence of the beam with distance.

Table 15-2 Recommended frequency and tolerance values of the ten-point ultrasound quality-control program

Performance evaluation	Frequency	Permissible tolerance
Sensitivity	Monthly	3 db
Distance calibration	Weekly	±2%
Caliper calibration	Monthly	±2%
Registration	Monthly	3 mm
Dead zone	Monthly	4 mm
Axial resolution	Weekly	1 mm
Lateral resolution	Monthly	3 mm
Time gain characteristic	Monthly	N/A
Display characteristic	Weekly	N/A
Visual inspection	Monthly	N/A

Review Questions: Chapter 15

1. Define or otherwise identify:
   a. Phantom
   b. Test object
   c. Sensitivity
   d. Dead zone
   e. Time gain compensation (TGC)

2. Describe the difference between quality assurance and quality control.

3. Describe the five sets of rods in the AIUM test object and the principle use of each.

4. The SUAR test object is designed to evaluate

what parameters? What is the advantage of a tissue equivalent phantom?

5. Identify two characteristics that should be evaluated weekly.

6. Which of the following has attenuation properties similar to soft tissue?
   a. AIUM 100 mm test object
   b. Beam profiler
   c. Tissue equivalent phantom
   d. Hydrometer

7. A quality assurance program
   a. is needed by all ultrasound departments.
   b. is necessary only in high volume departments.
   c. tests only transducer performance.
   d. is set up by all manufacturers.

8. Which of the following can be used in a quality assurance program?
   a. AIUM 100 mm test object
   b. Beam profiler
   c. Tissue equivalent transducer
   d. Artifact generator

9. It is necessary to check a machine's performance with various devices
   a. before every study.
   b. daily.
   c. only when you suspect a problem.
   d. routinely.

10. If pins in the AIUM test object do not appear in the same place on the display, what is the problem?
   a. Registration accuracy
   b. Range accuracy
   c. Near gain
   d. Gray-scale dynamic range

16 Biological Effects of Ultrasound

Our experience with diagnostic applications of x-radiation has made us especially sensitive when applying new energy modalities to patients. With the very earliest applications of x-radiation, it was recognized that high doses were related to harmful physiologic responses that could occur soon after the radiation exposure. For many decades, however, low doses of x-radiation, less than 25 rads (0.25 Gy), were considered to be totally safe. It was not until the 1930s that convincing evidence began to appear showing that even low doses could be responsible for harmful late effects.

Much investigation has been completed and more will continue to ensure that a similar experience does not occur regarding diagnostic ultrasound. At the present time, it can be said that diagnostic ultrasound produces absolutely no early physiologic responses in patients, and there is little likelihood of a late response following such exposure.

Present diagnostic pulse-echo ultrasound units produce SATA intensities of 5 to 20 mW cm^{-2} with SPTP intensities up to 10 W cm^{-2}. Doppler units produce average intensities of 10 to 30 mW cm^{-2}, and ultrasonic physiotherapy systems operate at SATA intensities of 1-5 W cm^{-2}.

One of the great difficulties with interpreting ultrasound bioeffects studies is dosimetry. **Ultrasound dosimetry** is very difficult to perform and in most studies is totally absent. To conduct ultrasound dosimetry one needs either a calibrated transducer, a hydrophone, or a radiation microbalance. The radiation microbalance is used for average power measurements and the hydrophone for mapping the spatial profile of the beam. Normally these devices are available only in very sophisticated laboratories. Consequently, although a precise knowledge of intensity, power, pulse width, repetition rate, spatial dimensions, and exposure duration are essential to proper-hazards evaluation, they are usually unavailable.

Because of the widespread use of diagnostic ultrasound, especially in obstetrics, radiobiologists will be kept busy for many years supplying supportive evidence for the safety of this modality. Since many such examinations occur at a time of particular sensitivity of the fetus to all environmental influences, it is absolutely essential that we understand the effects of ultrasound and the doses necessary to cause those effects. It is important that these investigations continue, because it is estimated that approximately one-half of all newborns in the United States are presently examined with ultrasound in utero, and the fraction is increasing.

MECHANISM OF ACTION

When investigating the harmful physiologic effects of diagnostic ultrasound, radiobiologists search for mechanisms of action. **A mechanism of action is the means by which the imaging field transfers energy to the target tissue.** In the case of x-radiation, the mechanism of action is ionization and excitation. In the case of diagnostic ultrasound, the mechanism of action appears to be either thermal or mechanical.

Thermal effects

As diagnostic ultrasound passes through tissue, molecules are caused to vibrate, as discussed in Chapter 1. This vibration results in an attenuation of the ultrasound beam and absorption of energy by the target tissue. The energy transfer from the ultrasound beam to the tissue results in an elevation of tissue temperature. The larger the specimen being irradiated, the larger the temperature rise, but it will take longer.

The higher the intensity of the ultrasound beam, the more vigorous the vibration of tissue molecules that results in higher tissue temperature. Temperature elevation is not an instantaneous phenomena because of the physiologic protective measures of the body.

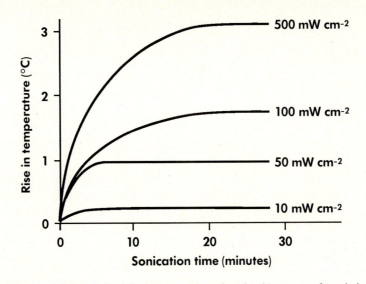

Fig. 16-1 Ultrasound induced physiologic temperature elevation in a mouse fetus is intensity and time dependent.

Thermal conduction from cell to cell and **thermal convection,** by way of body fluids, are examples of such measures. Fig. 16-1 shows the temperature rise in mouse tissue as a function of intensity and time. This effect is considerably less in humans because of the relative increase in mass to the ultrasound beam ratio.

A temperature rise of approximately 1°C (2.3°F) is required to be physiologically measurable. Normal body temperature is 36.7°C (98.6°F). In addition to diagnostic ultrasound, there are many other environmental agents that can result in temperature elevation. From experiments with many such agents we have obtained good dose-response data for generalized thermal effects. A temperature rise of several degrees Celsius in the whole body is known to be hazardous because of interference with normal homeostasis. However, when the heating is localized, as with diagnostic ultrasound, it is not considered significantly hazardous. The temperature rise resulting from a diagnostic application of ultrasound is about 0.1°C or less to the local tissue. Normal diurnal temperature variation can exceed 1°C, which is considered totally safe. During ultrasonic physiotherapy, localized temperature increases of 5°C during a 10 to 15 minute treatment are common. Fig. 16-2 represents the approximate ultrasound

intensity necessary to produce a 1°C temperature rise in tissue as a function of frequency. Under ideal conditions, sonication at approximately 300 mW cm^{-2} at 2 to 3 MHz would be required for a 1°C rise in temperature. The actual situation will depend on the size, composition, and vascularity of the subject tissue.

Potential thermal effects of diagnostic ultrasound are not considered a significant biologic hazard, mainly because the temperature elevation is so low. However, it is also known that any effect of an elevation of temperature at the tissue level is transient. Once the energy source is removed, the tissue temperature will return to normal with no residual effect.

Mechanical effects

If the ultrasound intensity is sufficiently high and the molecular structure of the target tissue sufficiently loose, the violent molecular agitations can result in small **microbubbles.** The production of such bubbles is termed **cavitation.**

Cavitation occurs when dissolved gases grow into bubbles around stable microbubble nuclei during the negative pressure phase of the propagated sound wave. Depending upon the frequency, a resonance can occur at a critical site inducing **microstreaming.** The microstreaming in turn produces regions of high

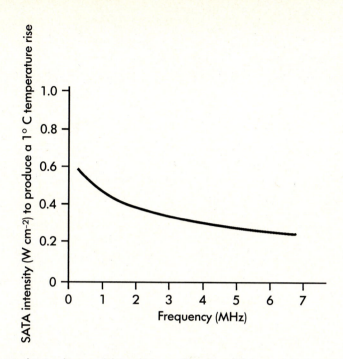

Fig. 16-2 Approximate ultrasound intensity as a function of frequency necessary to produce a 1° C rise in local tissue temperature.

shear that can result in molecular scission, the breakage of molecules. Intensities less than 1 W cm^{-2} will not produce cavitation in any tissue. Intensities greater than approximately 1000 W cm^{-2} are necessary for the formation of microbubbles. At such intensities the microbubbles can collapse during the compression phase of the ultrasound, resulting in violent molecular disruptions. Because such high intensities are required, it is clear that mechanical effects represent a remote potential for ultrasound damage.

Cavitation has been demonstrated in tissue-like matter subjected to very intense ultrasound fields but it has not been observed in-vivo. In order for cavitation to occur, dissolved gases must be present and there must be continuous wave ultrasound. **The very short pulse lengths associated with pulse-echo ultrasound will not produce cavitation** because there is not enough time to establish resonance. As with thermal effects, once the source of ultrasound is removed, the cavitation slowly resolves as the gases are again absorbed in the medium.

EFFECTS ON SIMPLE STRUCTURES

Although our interest lies in the effects of diagnostic ultrasound on humans, much has been gained by studying potential mechanisms of action through in-vitro studies. However, it is extremely difficult to extrapolate from the in-vitro situation to the clinical situation. The geometry is never the same and the in-vitro response may differ considerably from the in-vivo response because the mechanism of action may be different. Frequently, however, in-vitro experiments lead to new mechanisms and responses and suggest design criteria for in-vivo experiments. Such in-vitro studies generally fall into molecular or cellular investigations.

Molecular responses

There are two types of target molecules to be considered: water and macromolecules. Effects on such molecules have been observed, but only following extremely high intensities of ultrasound.

Water. The sheer forces produced by intense ultrasound in water can apparently produce free radical ions. A free radical ion is a molecule or radical with an unpaired electron in its outermost shell. This electronic configuration causes such a radical to be particularly reactive with organic molecules and can result in disruption of molecular bonds. The principal types of free radicals suspected of being formed are HO_2*, $H*$, and $OH*$. The mechanisms for such formation is as follows:

$$H_2O + U/S \rightarrow HOH^- \text{ or } HOH^+ \qquad (16\text{-}1)$$

The HOH^+ and HOH^- ions are relatively unstable and can dissociate into yet smaller molecules as follows:

$$HOH^+ \rightarrow H^+ + OH* \qquad (16\text{-}2)$$

$$HOH^- \rightarrow OH^- + H* \qquad (16\text{-}3)$$

$$H* + O_2 \rightarrow HO_2* \qquad (16\text{-}4)$$

Macromolecules. A macromolecular suspension such as a protein or a DNA conglomerate can be exceptionally viscous. The viscosity of such material can be dramatically reduced with intense ultrasound. This reduction in viscosity is caused by violent sheer forces that produce a scission of the main backbone of the macromolecule. Such main-chain scission represents a sufficiently violent molecular response that, were it to occur at diagnostic levels of ultrasound, harmful physiologic effects might be anticipated.

Chromosomes. Chromosomes consist of conjugated macromolecules—proteins, DNA, and water. Our extensive knowledge of the effects of ionizing radiation on chromosomes makes them perfect objects for ultrasound investigations. In the 1960s a few investigations with human lymphocytes, irradiated in vitro, reported chromosome fragmentation and translocation as observed biologic responses. Many studies since then have failed to corroborate such findings, even though a wide spectrum of ultrasound devices and intensities have been employed. Both continuous wave and pulse-echo ultrasound, at SATA intensities up to 10 W cm^{-2} for irradiation times to 24 hours, have failed to show chromosome damage.

Cells

Single cell survival experiments, similar to those of ionizing radiation, have failed to show a response. Differences between sonicated cells and controls are nil. Furthermore, it seems that cells and therefore tissues do not experience cumulative effects.

OBSERVATIONS ON ANIMALS

Many studies on animals, especially mice, rats, chickens, and rabbits have been conducted. Experiments with mature animals have consistently been negative at intensity levels less than 20 W cm^{-2}. Experiments with pregnant animals have elicited some responses at intensity levels considerably above those employed for diagnostic ultrasound.

Following irradiation during pregnancy, some studies have shown a retardation of growth as the principle response. These experiments are difficult to interpret because the degree of retardation does not seem to be related to the time or the intensity of the ultrasound irradiation. Some studies have been able to induce congenital abnormalities, but such a response is equivocal because of the inability to replicate the experiments. There are also some suggestions of a neurologic deficit following sonication in-utero. Shortly after birth, subject newborns are not as capable of providing a conditioned response as controls. Again, this finding is equivocal. Paraplegia and hemorrhage into the cord has been reported in adult rats at intensities of 50 W cm^{-2}. The response was related to frequency, but again, this is orders of magnitude higher than diagnostic levels.

None of these responses has been shown to occur with pulsed ultrasound. The positive findings resulted from studies using continuous-wave ultrasound of considerable duration stretching to hours. Intensities of at least 10 W cm^{-2} were required. The statement on mammalian in-vivo biological effects by the American Institute of Ultrasound in Medicine (AIUM) says it all:

> In the low megahertz frequency range there have been no independently confirmed significant biological effects in mammalian tissues exposed to intensities below 100 mW cm^{-2}. Furthermore, for ultrasonic exposure times less than 500 s and greater then 1 s, such effects have not been demonstrated even at higher intensities, when the product of intensity and exposure time is less than 50 J cm^{-2}.

This statement applies to ultrasound frequencies from 0.5 to 10 MHz, the range employed for imaging. The intensity referred to is the SPTA intensity.

When interpreting mammalian in-vivo data, several conditions should be recognized. Although nearly all studies used continuous-wave ultrasound, there was a good mix of both focused and unfocused beams. All such observations were on animals. The proper extrapolation to man is unknown for two principle reasons. First, the animal fetus is much smaller than the human fetus relative to the size of the ultrasonic beam and the number of wavelengths in the fetus. Second, although we may be able to properly characterize the ultrasound beam in water, sonication of a small experimental animal distorts the field so much that the actual fetal intensity may be greatly different from that expected. Very little research has been done at diagnostic levels because of the inability to elicit a response.

OBSERVATIONS ON HUMANS

There are two human populations, diagnostic patients and therapy patients, on which intensive observations have been made. Various therapeutic applications of ultrasound have been made for some 30 years now. Patients who have received such therapeutic intensities of ultrasound have been the subject of many scientific observations. The widespread application of diagnostic ultrasound is more recent. Diagnostic ultrasound was introduced into obstetric practice in 1966, reaching common clinical use by about 1975. Many retrospective studies of diagnostic patients have been conducted, the results of which form the principle basis for current biosafety recommendations.

Therapy patients

There are many therapeutic applications of ultrasound where a biologic response is deliberately sought. None, however, are supported in principle by basic science. Nevertheless, it is from therapeutic applications that we have our only reliable, human data. Therapeutic ultrasound is used to assist in bone and joint healing. It is also used in sports medicine to promote muscle relaxation as in diathermy. It has been employed surgically to cut and coalesce tissues, and it is currently employed in nearly every dental office as a scalpel to assist in cleaning teeth.

Physical therapy. A typical therapeutic procedure for treatment of arthritis, bursitis, and to reduce swelling of strained and inflamed joints is to apply high-intensity continuous-wave ultrasound to the affected tissue. A high-intensity, wide-area ultrasound beam is positioned over the inflamed joint for 3 to 10 minutes. During treatment the transducer might be moved or manipulated to ensure heating of all intraarticular tissues.

A similar procedure is used with ultrasonic diathermy. Such an application of ultrasound apparently stimulates dilation of peripheral vessels and thereby improves blood perfusion. The effect is the same as using a heating pad, except deeper and selective

tissues are involved. Depending upon the intensity of ultrasound, treatments lasting up to 30 minutes may be administered.

Hyperthermia. A recent and promising application of high-intensity ultrasound is hyperthermia. Apparently at about 42°C an inhibition of metabolism in malignant cells is evident. Normal cells are unaffected. When ionizing radiation is applied simultaneously, or near the time of sonication, the effect is amplified.

Surgery. Surgical applications of medical ultrasound are rare, but they have been available for many years. In such a procedure, a very high-powered ultrasound transducer is focused to the tissue of interest. The ultrasound intensity is extremely high, from 50 to 100 W cm^{-2}, and the application time is in seconds rather than minutes. There is disintegration of tissue along the path of the focused beam. Most such applications have not been adopted because alternative techniques always prove superior. An exception is the exploding use of extracorporal ultrasonic lithotripsy for removal of renal and biliary stones.

Dentistry. We are all familiar with the ultrasonic scalpel that is used to descale teeth. The tip of the scalpel is caused to vibrate ultrasonically in the 20 to 40 kHz range. It is, in effect, a miniature jackhammer. Water is simultaneously diverted over the tip of the ultrasonic scalpel for cooling.

Ultrasound therapy patients now measure in the millions. In the many published studies on such patients, none have reported harmful effects. It is often not known what the ultrasound intensity was, but surely it was in the 10 to 100 W cm^{-2} range. For some applications in certain clinics, the procedure was to increase the intensity of the ultrasound applicator until the patient senses pain. At that instance, the intensity would then be reduced to just below the pain threshold, and the treatment would proceed. Although an immediate sensation may be reported by the patient, no lasting effects have ever been noted.

Diagnostic patients

Many studies have been reported that involve small populations measuring only hundreds of patients. Most such studies are simply anecdotal in nature. On the other hand, there are several large scale, epidemiologic investigations on patients subjected to diagnostic ultrasound. Some of these investigations involve over 100,000 patients, one involving 1.2

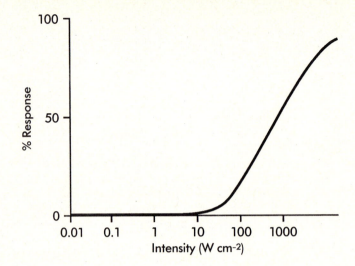

Fig. 16-3 Ultrasound bioeffects most likely follow a linear, threshold type dose response relationship.

million examinations, and deal with obstetric applications of both pulse-echo and continuous-wave diagnostic ultrasound. The experimental plan was usually designed to detect any measure of congenital malformation, spontaneous abortion, birth weight, premature labor, and chromosome aberrations. The reported results to date are consistently negative. Occasionally, a report will appear suggesting a positive response, but in each case attempts by other investigators to demonstrate the same response have failed.

Regardless of the absence of any evidence of harmful effects on diagnostic patients, physicians have and will continue to use ultrasound with discretion. Presently approximately one half of all newborns are examined with ultrasound in utero; consequently it becomes more difficult to identify control populations. This situation will keep radiobiologists busy for many years as they try to identify more subtle effects, such as biochemical aberrations and minor behavioral changes.

A summary of representative intensity levels employed in both therapy and diagnosis is given in Table 16-1.

DOSE-RESPONSE RELATIONSHIPS

Nearly all radiobiologic investigations are designed to develop dose-response relationships, and inves-

Table 16-1 Representative intensities for various applications of medical ultrasound

Application	Intensity range (SATA)
Surgery	>10 W cm^{-2}
Therapy	0.5-3 W cm^{-2}
Diagnostic	
Perivascular Doppler	50-500 mW cm^{-2}
Other	1-50 mW cm^{-2}

tigations with ultrasound are no exception. There are two principle reasons to construct such dose-response relationships. The first is to predict the therapeutic value of a given ultrasound intensity. The second is to be able to assess any potential hazards following various levels of diagnostic ultrasound.

An analysis of the total radiobiologic data shows that the dose-response relationship is threshold in form. Depending upon the nature of the response being observed, the threshold is somewhere in the neighborhood of 10 W cm^{-2}. The shape of the dose-response relationship above the threshold dose is unknown, but in all likelihood, it is sigmoid. Such a non-linear, threshold dose-response relationship is shown in Fig. 16-3.

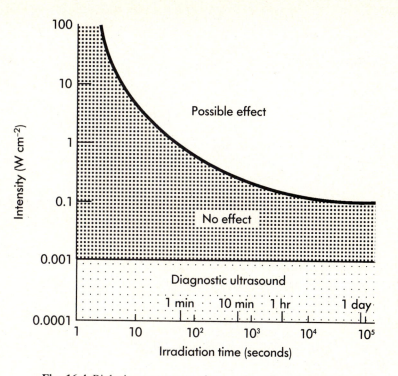

Fig. 16-4 Biologic responses to ultrasound are time dependent.

As with most other physical agents, responses to ultrasound are modified by the time over which the ultrasound is administered. As a modifying factor, time is very significant when considering the bioeffects of ultrasound. A summary relationship is that shown in Fig. 16-4. Here it is seen that regardless of the time of application, if the intensity does not exceed approximately 10 W cm^{-2}, there will be no response. Following exposure times less than approximately 10 minutes, intensities in excess of 100 W cm^{-2} may be required to elicit a response.

CLINICAL SAFETY

The most important reason for even considering such radiobiologic factors is to provide assurance that diagnostic ultrasound is being applied safely. In addition to acoustical safety, clinical safety really involves potential electrical/mechanical hazards. Both require that the ultrasonographer understand the instrumentation sufficiently to ensure against any patient injury. Therefore a knowledge of the electrical,

mechanical, acoustical, and operating parameters is essential.

Acoustical safety is perhaps most important since reports of electrical or mechanical injury are nonexistent. Of course, substantial reports of acoustical injury are also nonexistent. In view of the foregoing discussion, it is clear that diagnostic levels of ultrasound do not pose a hazard to the patient or the ultrasonographer. Because of the great benefits of diagnostic ultrasound to patients, the statement of the American Institute of Ultrasound in Medicine (AIUM) should be accepted by all:

No confirmed biological effects on patients or instrument operators caused by exposure at intensities typical of present diagnostic ultrasound instruments have ever been reported. Although the possibility exists that such biological effects may be identified in the future, current data indicate that the benefits to patients of the prudent use of diagnostic ultrasound outweigh the risks, if any, that may be present.

Still as an energy agent, its use should be conditioned with caution. The ultrasonographer should know the operating characteristics of the instrument and apply the lowest practical acoustic intensity in the shortest possible time to provide the required diagnostic information. Patient dose can be lowered by reducing the gain and examination time. The scanner manufacturer aids the ultrasonography by producing a relatively low transmitter output and the highest practical receiver gain. Furthermore, the time of the investigation should be as short as possible, and a routine quality assurance program should be in use. The recent recommendations of the American Institute of Ultrasound Medicine (AIUM) should be adhered to.

The intentional application of diagnostic ultrasound without benefit of medical information should be avoided. The use of live models and sales personnel to demonstrate diagnostic ultrasound devices must be prohibited.

Therapeutic ultrasound should not be applied to sensitive areas of a pregnant or potentially pregnant patient. The 10-day rule developed for x-ray imaging should be instituted for ultrasound therapy treatment of the abdomen, pelvis, or lower back.

Other conditions of therapy should be avoided, such as exposure of eyes, testes, heart, growth plates of pediatric bones, and any other tissue with diminished vascularity.

Review Questions: Chapter 16

1. Define or otherwise identify:
   a. Hydrophone
   b. Mechanism of action
   c. Cavitation
   d. Free radical ion
   e. Diathermy
   f. Hypothermia

2. What is the range of SATA intensities employed with diagnostic imaging systems?

3. What fraction of pregnancies are examined by diagnostic ultrasound?

4. Discuss the elevation of temperature as a consequence of ultrasonic examination.

5. What are the three principal mechanical effects of ultrasound?

6. What is the shape of the dose response relationship for diagnostic ultrasound?

7. Discuss the hazard involved with a diagnostic ultrasound examination.

Appendix

Answers: Chapter 2

1. a. 0.7
   c. 0.00004
   e. 0.375
2. a. 5.63×10^9
   b. 9.2×10^{-7}
   c. 9.41×10^2
   d. 2.53×10^{-1}
   e. 2.5×10^{-4}
   f. 3.33×10^3
3. a. $x = 5$
   b. $x = 3.75$
   c. $x = 1.86$
   d. $x = 7.35$
   e. $x = 3TV/BQ$
4. a. 1.4×10^5 meters
   b. 7.5×10^{-5} volts
   c. 3.0×10^{-1} seconds
   d. 6.05×10^{-2} grams
5. a. 7.0 MHz
   b. 1.6 ms
   c. 180 μg
   d. 44 km
   e. 4 mm
6. a. 6
   c. 29
   e. 10010
8. Weigh your shoe in pounds and convert to metric; measure length in inches and cm.

Answers: Chapter 3

2. 1500 m s^{-1}
3. 0.25 μs
4. 6.5 μs
5. The tumor will appear closer since the machine measures time, not distance.
6. 6.76 J, 22.5 Watts
7. $1{,}000$ kg m^{-3}
8. 10^5 dynes

9. 10^7 ergs
11. a, c, d, b, e, f
12. Discuss the mechanical equivalent of heat.

Answers: Chapter 4

3. b. 1 m
   d. 0.33 seconds
   f. 3 m s^{-1}
4. 3.08×10^{-4} m, .308 mm
6. b, c, a, e, d
8. a. $v = f\lambda$
   $$= (4 \times 10^6 \ 1/s)(1 \times 10^{-3} \ m)$$
   $$= (4 \times 10^3 \ m \ s^{-1})$$
   b. $f = v/\lambda = \dfrac{4000 \ m \ s^{-1}}{3 \times 10^{-3} \ m} = 1.33$ MHz
   c. 1.33 MHz

Answers: Chapter 5

3. Focus the beam to a smaller area and increase the power level.
5. 6 mW cm^{-2}
6. Intensity is proportional to the amplitude squared.
7. -40 dB
8. -30 dB, 99.9%, .1%
10. c. SATP and SATA; SPTP and SPTA

Answers: Chapter 6

3. Refraction
4. 20 rayls
5. $Z_{air} = 0.0004$ rayls, the acoustic impedance of other materials is much higher.
6. % R $= 4\%$, % T $= 100 - 4 = 96\%$
8. % R $= 48.9\%$
9. a. Increasing density increases Z; percent R increased.

b. No change in reflection, however beam attenuation increases.
c. Percent R will decrease.
d. No change.
e. No change.
11. 2 MHz × 1 dB cm^{-1} MHz^{-1} × 5 cm = 10 dB. 10 dB is 90% attenuation so the remaining intensity is 20 mW cm^{-2} × 0.1 = 2 mW cm^{-2}.

Answers: Chapter 7

2. Loudspeaker and microphone
5. Quartz lithium niobate, lithium sulfate, lead zirconate titanate, barium titanate, lead metanyobate
6. Meters per volt (m v^{-1})
8. Its thickness which should be half the wavelength of the ultrasound employed. Thickness determines frequency.
9. 0.4 mm
10. The acoustic impedance should have a value of between 10 to 20 rayls, and the matching layer should be ¼ wavelength of the ultrasound beam.
11. 2.27

Answers: Chapter 8

1. a. the number of pulses emitted each second
   b. the time in seconds from the beginning of one pulse to the beginning of the next
   c. the time during which the pulse actually occurs
   d. the fraction of time the ultrasound is actually being emitted
   e. the length over which an ultrasound pulse occurs
   f. emission with equal intensity in all directions
2. a. 20 Hz
   c. 50 ms
   d. 2 ms
3. a. 2 ms
   b. 1 μs
   c. .0005
   d. 1.54 mm
4. 13.5 cm
6. 0.96°
8. a. increases
   b. no change
9. a. false

b. true
c. true

Answers: Chapter 9

2. System noise
3. The shorter the SPL, the better the axial resolution. The higher the ultrasound frequency, the better the axial resolution. Damping improves axial resolution.
4. 0.77 mm
6. Because the beam diameter varies with depth
7. a. true
   b. false
   c. true
   f. false
9. 0.5 cm

Answers: Chapter 10

2. 7.2 cm
5. Echoes are indicated by spikes on the monitor; the height of the spike is proportional to the amplitude of the returning echo.
6. a. 2.3 cm

Answers: Chapter 11

2. It's quicker, easier, and possibly portable. The image is continuously refreshed, and imaging of movement of internal structures is possible.
3. 20 frames per second

Answers: Chapter 12

3. Compression of a large range of amplitudes from the receiver into a range that can be handled by other system components
5. Digital memory
7. TGC
8. See the binary number system in Chapter 2.

Answers: Chapter 13

2. Moving structures scatter the ultrasound beam back into the transducer. Since the scatterers are

moving, the frequency of the ultrasound beam returning is slightly different from that emitted. The machine may use this Doppler-shifted frequency to determine the magnitude and direction of the motion if the Doppler angle is known.

5. a. 1300 Hz

   b. 5.0013 MHz

6. Cos 45° = .707

8. Lack of range resolution, use pulsed Doppler, no aliasing.

9. Doppler systems provide information only on moving structures. Stationary objects do not alter the frequency of the returning signal. Thus the Doppler instrument cannot distinguish stationary objects and cannot therefore provide an anatomical image. Many duplex Doppler systems alternate between real-time B-mode and Doppler, and provide the capability of viewing flow and motion information superimposed on the pulse-echo image.

Answers: Chapter 14

1. a. any unintended information on an image that does not represent the object

   b. multiple reflection of an ultrasound beam between two interfaces

   c. giving the appearance of a mass when there really is not one

   d. reflection from a smooth interface

   e. misalignment of compound B-scan traces with movement of the articulated arm

   f. an artifact produced by insufficient sampling

2. Reverberation, shadowing, displacement, distortion, aliasing

3. Reverberation

4. Reflective attenuation and edge shadows—reflective shadows are more easily produced and bothersome.

5. Misregistration, multipath, beam width

6. Doppler imaging

Answers: Chapter 15

2. Quality control assesses the adequacy of instrumentation and equipment. It is a component of quality assurance that deals with the overall management of the patient.

3. Group A—vertical row—used for depth calibration, linearity, and gain.

   Group B—bottom row—used for sensitivity, horizontal calibration, and horizontal linearity.

   Group C—vertical row—used for lateral resolution and beam width.

   Group D—horizontal row—used to evaluate dead zone.

   Group E—central grouping to evaluate axial resolution and registration.

4. Sensitivity, uniformity, and axial resolution

5. Distance calibration, axial resolution, display characteristics.

8. a

10. a

Answers: Chapter 16

2. 5 to 30 mW cm^{-2}

3. Approximately half

4. A 1° temperature rise is necessary to be physiologically significant. Diagnostic ultrasound increases temperature 0.1°C or less.

5. Cavitation, microstreaming, molecular scission

6. Threshold and non-linear

7. The risk of injury is essentially zero.

Index

A

Absorption in attenuation of ultrasound beam, 42-43
Acceleration, 15
Acceptance-test images, 117
Acoustic coupling, 56
Acoustic impedance, 47-49
  of transducer element, 53
Acoustic safety, 161
Acoustic variables
  amplitude of, 35
  and ultrasonic intensity, 33
Acoustic waves, characteristics of, 24
Activation
  segmental, 93-95
  sequential, 93, 94
ADC; *see* Analog-to-digital converter
Addresses in analog-to-digital converter, 110
Air, interface of, with tissue, reflection at, 47
Air bubble, artifact due to, 134
AIUM test object, 143-144, 145
  routine measurements and observations with, 148, 153
Algebra, review of, 6
Aliasing, 127-128, 141, 142
  as artifact on spectral display, 130
American Institute of Ultrasound in Medicine
  statements of, concerning biological effects, 158, 161, 162
  test object of, 143-144, 145
    routine measurements and observations with, 148, 153
A-mode; *see* Amplitude mode
Amplification in image processing, 105-107
Amplitude of acoustic variables, 35
Amplitude mode, 78, 79-82
  calibration of instruments for, 79
  transducers for, 79, 81
Analog scan converter, 86, 108, 110
Analog-to-digital converter (ADC), 110
Angle
  Doppler, 121-123
  of incidence in reflection, 45-47
  sector, in real-time transducer, 91
Annular array(s), 97
  advantages and disadvantages of, 99

Annular array(s)—cont'd
  electronic focusing of, 101
Anti-node in acoustic wave, 24
Apodization, 101
AR; *see* Axial resolution
Arc scan, 85
Array(s), 62
  annular, 97
  convex, 95, 96
  electronic real-time, 93-97
Artifact(s)
  aliasing, on spectral display, 130
  definition of, 133
  image anaylsis, 133
  refraction, 44
Asimuthal resolution, 73
Attenuation, 40-41
  example of, 49
  interactions of, 42-44
  and time-gain compensation, 106-107
Attenuation coefficients, 40
Attenuation shadows, 137
Axes of graphs, 10, 11
Axial resolution, 70-73
  computing, 73
  in quality control, 150-151
  SUAR test object and, 146

B

Back-scatter, 44
Bandwidth, 56-57
Barium titanate, 3
  in transducer element, 53
Base quantities, 10, 11
Bats, ultrasound use by, 1
Beam, ultrasound, 59-68
  characteristics of pulses in, 59-61
  focusing, 65-67
  near and far field, 62-65
  percentage reflected, 47
  reflection of, 36
  spatial and temporal variation in, 37-38
  steering, 97, 102-103
  wave front in, 61-62
Beam angle, incident, effect of, on Doppler shift frequency, 121-123

Beam overlap region, 124
Beam width
    artifacts related to, 140-141
    and lateral resolution, 75, 76
Bell, Alexander Graham, 35
Binary digit, 9, 10
Binary numbers, 9
    conversion of, to decimals, in scan converter, 110, 111
    review of, 8-10
Binary word length, 110, 111
Biochemistry, 18
Biological materials
    attenuation coefficients in, 40
Bistable images, 85-86, 111
Bit, 9, 10
Black and white display mode, 85-86, 111
Bliss, W.R., 2
Blood cells, Doppler imaging of, 128
Blood flow, Doppler angle and, 121
Blood vessels, Doppler imaging of, 128-130
B-mode; *see* Brightness mode
Bone, attenuation shadows from, 137
Bowel, gas-filled, artifact from, 133, 136
Breast cyst, enhancement artifact and, 139
Brightness mode, 78, 82-87
    compound, 82
    development of scanner for, 2-3
    real-time, and Doppler transducers in one probe, 130
    static, misregistration artifact in, 139
British system of units, 13
Bytes, 9, 10

C

Calibration
    caliper, 149
    distance, in quality control, 148-149
    of instruments for amplitude mode, 79
Caliper calibration, 149
Calorie, definition of, 19
Camera(s)
    gray-scale capacity of, in quality control, 151
    multiformat, 116-117
Cardiac cycle, Doppler analysis of, 128-129
Cardiography, ultrasonic, 87
Carotid artery
    aliasing in image of, 141, 142
    Doppler spectrum of, 128
Case of single element transducer, 53
Cathode ray tube (CRT), 113
Cavitation, 156-157

Cells, biological effects of ultrasound on, 158
Celsius temperature scale, 19, 20
Ceramics in transducer element, 53
CGS system of units, 13
Chemical energy, 18
Chromosomes, ultrasound effects on, 158
Clinical safety, 161-162
Clip, surgical, artifact from, 135
Coefficients, attenuation, 40
Color flow Doppler systems, 130-131
Comet tail artifact, 133-134, 135
Compound B-mode, 82
Compound scan, 85
Compressibility and density, effects of, on velocity, 27-28
Compression
    in acoustic wave, 24
    in receiver, 105
Computed tomography
    aliasing in, 142
    resolution characteristics of, 69
Computing lateral resolution, 76
Conduction of heat, 19, 156
Conservation of energy, law of, 18
Continuous wave ultrasound, 78
    biological effects of, 158
    Doppler, 123-124, 125
    duty factor of, 61
    frequency of, 56
    in physical therapy, 159
Contrast resolution, 69
Contrast-detail curve, 69, 70
Convection of heat, 19, 156
Conversion
    from binary to decimal numbers in scan converter, 110, 111
    scan, 108-113
Converter, scan, 86
    analog, 108, 110
    digital, 110-113
Convex array probe, 95, 96
Coordinates in graphs, 10
Coronal image of fetus, artifact on, 136, 137
Coupling, acoustic, 56
Coupling factor, electromechanical, 53
Critical angle of gallbladder, 137, 138
CRT; *see* Cathode ray tube
Crystals(s), piezoelectric, 51, 52
    for continuous wave Doppler, 123-124
    receiving, 105
    in single element transducer, 53-54

Curie, Jacques and Pierre, 1
CW Doppler; *see* Continuous wave Doppler
Cysts, enhancement artifact and, 137, 139

D

DAC; *see* Digital-to-analog converter
Damping factor in axial resolution, 73
Damping material in single element transducer, 54
Dead zone, evaluation of, 149-150
Decibels, 35-36
   versus intensity of reflected wave, 37
Decimals
   conversion of binary numbers to, in scan converter, 110, 111
   review of, 5
Definition
   image, and line density, 89
   in radiology, 69
Delay control, 107
Delay times, focusing and, 103
Demodulation, 124
   in signal processing, 107-108
Density and compressibility, effects of, on velocity, 27-28
Dentistry, 159
Depth, 70
   of image, maximum, in real-time transducer, 91
Depth penetration and frequency, 73
Depth-gain compensation (DGC), 106-108
Derived quantities, 10, 11
Destructive interference, 31
Detail in radiology, 69
Detection, signal, 108, 109
DF; *see* Duty factor
DGC; *see* Depth-gain compensation
Diathermy, ultrasonic, 159
Digital scan converter(s), 86, 110-113
   beam-intensity artifact and, 141-142
   microprocessor and, 88
Digital-to-analog converter (DAC), 113, 114
Dipoles in piezoelectric crystals, 51, 52
Direction of reflection, 79
Displacement as artifact, 138-141
Display devices, 113-116
   evaluation of, 151
Display modes, 85-86
Distance calibration in quality control, 148-149
Distortion artifacts, 141-142
   in radiology, 69
Divergence of far field, 64-66
Dog whistle, 1, 2

Doppler, Christian, 119
Doppler angle, 121-123
Doppler equation, 119-121
Doppler flow measurements, 78
Doppler shift frequency (f_D), 119
   effect of incident beam angle on, 121-123
   Nyquist limit and, 127-128
   for various Doppler angles and interface speeds, 123
Doppler ultrasound, 119-132
   aliasing and, 127-128
   color flow, 130-131
   continuous wave, 123-124, 125
   duplex, 130
   mathematics in, 119-121
   pulsed, 124-127
   in spectral analysis, 128-130
Dose-response relationships, 160-161
Dosimetry, ultrasound, 155
Dual focusing in linear arrays, 99
Duplex Doppler, 130
Dussik, K.T., 2
Duty factor (DF), 61
Dynamic electronic focusing, 102
Dynamic modes, 78
Dynamic range of receiver, 105

E

Edge shadows, 137, 138
Electric field, generated, 53
Electric potentiometers, 84
Electrical energy, 18
Electromagnetic radiation
   and sound/ultrasound, comparison of, 23
   velocity of, 26
Electromagnetic waves, 24
Electromechanical coupling factor, 53
Electron beam in cathode ray tube, 113-115
Electronic calipers in quality control, 149
Electronic focusing, 99-101
   dynamic, 102
Encoders, optical, 84
Encoding into binary digits, 10
Energy, 18-20
   law of conservation of, 18
   transfer of, from ultrasound beam to tissue, 155
Engineering prefixes, 14
Enhancement, 107
   of image, depth of penetration and, 137-138, 139
Envelope, video, after smoothing, 108
Enveloping in signal processing, 108

Equation
    Doppler, 119-121
    range, 78
    wave, 26-27
Equipment-performance monitoring, periodic, 143
Exponents, review of, 8, 9
Extracorporal ultrasonic lithotripsy, 159

F

f_D; *see* Doppler shift frequency
Fahrenheit temperature scale, 19, 20
Far field of ultrasound beam, 62-65
Far gain, 107
Fast Fourier Transform, 128
Fat layers, abdominal, reverberation and, 133
Fetal heart, monitoring of, 124
Fetus
    ability to observe motion of, 88
    artifact on image of, 136, 137
    sensitivity of, to ultrasound, 155
Fields, video, 115
Film processor, 117
First law of motion, 15
Flickering of image, 115
Flow imaging, 88
Fluid paths, short and long, 92
Fluid-filled tissues, beam-width artifacts and, 141
Fluid-tissue interface and reverberation artifacts,
        133
Focusing of ultrasound beam, 65-67
    in real-time ultrasound, 98-102
Force, Newton's second law and, 15
FR; *see* Frame rate
Fractions, review of, 5
Frame rate (FR)
    of electronic real-time transducers, 93
    and phased array, 97
    in real-time imaging, 90
Fraunhofer zone, 63
Free radical ion, 158
Frequency(ies)
    in attenuation of ultrasound beam, 43
    axial resolution and, 70, 72-73
    description of, 25-26
    and lateral resolution, 75, 77
    resonant, 56
    of ultrasound, 1
Fresnel zone, 63

G

Gain, 105-107

Gallbladder
    critical angle of, 137, 138
    sludge in, as artifact, 141
Gallstone, artifact from, 136, 137
Galton, Sir Francis, 1
Gas-filled bowel, artifact from, 133, 136
Gate width, 126
Gel, acoustic coupling, 56
Generated electric field, 53
Geometric distortion, 141
Graphing, review of, 10, 11
Grating, range, 126
Gravity, law of, 5
Gray scale
    assignment of, by scan converter, 108
    camera, in quality control, 151
    matrices required for, 112, 113
    postprocessing effects on, 114
Gray scale display mode, 85-86

H

Hardcopy, comparison of, with CRT image, 151
Heart, fetal, monitoring of, 124
Heat, 19-20
    mechanical equivalent of, 20
Heat energy, 18-19
Hertz, Heinrick, 25
High frequency multi-plane probes, 98
History of ultrasound, 1-4
Horizontal axis, 10
Howry, Douglass, 2-3
Humans, ultrasound effects on, 159-160
Huygen, Christen, 31
Huygen's principle, 30, 31
Hyperthermia, 159

I

Image
    analysis of, 133-142
        displacement in, 138-141
        distortion in, 141-142
        enhancement in, 137-138, 139
        equipment-performance monitoring based on, 143
        reverberation in, 133-134, 135
        shadowing in, 134-137
    bistable, 85-86
    definition of, and line density, 89
    maximum depth of, in real-time transducer, 91
Image quality GET FR Quality, 101
    quality of, and lateral resolution, 73
    recording of, 116-117

Image processing and display, 105-108
  display devices in, 113-116
  image recording in, 116-117
  preprocessing and postprocessing in, 113, 114
  pulser in, 105
  receiver in, 105-108
  scan conversion in, 108-113
Imagers
  real-time, types of, 89
Imaging
  flow, 88
  relationships among characteristics of, 91-92
Impedance, acoustic, 47-49
  of transducer element, 53
Incidence, angle of, in reflection, 45-47
Incident beam angle, effect of, on Doppler shift
      frequency, 121-123
Industrial applications of ultrasound, 2
Inertia, law of, 15, 16
Infrasound, 1, 2
Inspection, visual, of scanner, 151-152
Integration in signal processing, 108, 109
Intensity
  for medical uses of ultrasound, 160
  of reflected echoes, 79
  specification of, 36-38
  of ultrasound pulse, pulser and, 105
  versus decibels of reflected wave, 37
Interface, reflective, and reverberation artifacts, 133
Interference, 45
  destructive, 31
Interlaced video fields, 115
Internal motion, ability to observe, 88

J

JCAHO; *see* Joint Commission on Accreditation of
      Healthcare Organizations
Joint Commission on Accreditation of Healthcare
      Organizations (JCAHO), 143
Joule, James, 20

K

Kelvin temperature scale, 19, 20
Kilogram, standard, 12
Kinetic energy, 18

L

Laser printers, 116
Lateral resolution, 65-66, 73-77
  computing, 76
  in quality control, 150-151

Law(s)
  of conservation of energy, 18
  of gravity, 5
  of motion, Newton's, 15-16
  Snell's, 43, 137
LD; *see* Line density
Lead metaniobate in transducer element, 53
Lead zirconate titanate, 3
Leading edge detection, 108, 109
Left portal vein edge shadow, 138, 139
Length, standard of, 12
Lenses, focusing transducer with, 66
Light, speed of, 8
Line density (LD)
  of electronic real-time transducers, 93
  and phased array, 97
  of real-time transducers, 88-90
Line number of real-time transducers, 90
Linear array of real-time transducers, 93-96
Linear array transducers, electronic focusing of, 99
Linear potentiometers, 84
Linear scan, 85
"Listening window," 126
Lithium niobate in transducer element, 53
Lithium sulfate in transducer element, 53
Lithium sulphatemonohydrate, 3
Lithotripsy, extracorporal ultrasonic, 159
Liver, cystic lesions in, 137
LN; *see* Line number
Logarithmic compression, 105
Logarithms, 36, 37
Longitudinal resolution, 70
Longitudinal waves, characteristics of, 24

M

Macromolecules, ultrasound effects on, 158
Magnetic resonance imaging
  aliasing in, 142
  resolution characteristics of, 69
Mass, standard of, 12
Matching layer in single element transducer, 54-56
Mathematical background for physics, 5-10
Mathematics, Doppler, 119-123
Matrix, dielectric, in analog scan converter, 108
Maximum depth of image in real-time transducer, 91
Measurement(s)
  routine, in quality control, 148-152
  units of, 10-13
Mechanical effects of ultrasound on tissue, 156-157
Mechanical energy, 18
Mechanical equivalent of heat, 20

Mechanical real-time transducers, 92-93
    advantages and disadvantages of, 99
Mechanical waves, 21-24
Medical applications of ultrasound, 2-4
Memory
    personal computer, 10
    scan converter as, 108
Memory locations in analog-to-digital converter,
        110
Memory resolution, 111
Meter, standard, 12
Microbubbles, 156
Microprocessor
    and digital scan converter, 88
    and steering ultrasound beam, 103
Microstreaming, 156
Mineral oil for acoustic coupling, 56
Misregistration artifact, 139
MKS system of units, 13
Molecular scission, 157
Monitors, television, 113
Motion
    internal, ability to observe, 88
    Newton's laws of, 15-16
Motion mode, 87
Multiformat cameras, 116-117
Multipath artifacts, 139-140
Multi-plane probes, high frequency, 98

N

Near field of ultrasound beam, 62-65
Near gain, 107
Neovascularity, tumor, evaluation of, 131
Newton, Sir Isaac, 15
Newton's laws of motion, 15-16
Nodes in acoustic wave, 24
Noise, 69
Non-directional systems, 124
Nonfocused transducers, near field lengths and far field
        divergence of, 66
Non-specular reflection, 44
Notation, scientific, review of, 7-8
Nuclear energy, 19
Number system, binary, review of, 8-10
Nyquist limit, 127-128

O

Observations, routine, in quality control, 148-152
Obstetrics, probe for, 98
Optical encoders, 84
Organ perfusion, evaluation of, 131

Oscillating imagers, 92
Oscillation of particles in waves, 21

P

PA; see Pulse average
Particles, oscillation of, in waves, 21
Particulate matter in gallbladder as artifact, 141
Paths, fluid, short and long, 92
PD; see Pulse duration
Peak
    in acoustic wave, 24
    detection of, 108, 109
Peak velocities, aliasing at, 141
Pelvic examination, probe for, 98
Percentage of beam reflected (%R), 47
Percentage of beam transmitted (%R), 47
Perfusion, organ, evaluation of, 131
Period, description of, 26
Periodic equipment-performance monitoring, 143
Peristalsis, ability to observe, 88
Personal computers, 10
Phantoms, 143
    tissue, RMI, 146-148
Phase of wave, 28-30
Phased array, 96, 97
Phased-array transducers
    advantages and disadvantages of, 99
    for pulsed Doppler, 127
Physical therapy, ultrasound, effects of, 159
Physics, fundamentals of, 5-14
    mathematical background of, 5-10
    scientific prefixes in, 13-14
    units of measurement in, 10-13
Physiotherapy, ultrasonic, 156
Piezoelectric effect
    discovery of, 1
    in ultrasound transducer, 51-52
Pixel, 113
Plane wave front, 61-63
Plastic lenses, focusing transducer with, 66
PM-mode; see Position-motion mode
Polaroid image recording, 116
Portal vein(s)
    false, as multipath artifacts, 140
    left, edge shadow, 138, 139
Position-motion mode, 87
Postprocessing, 113
Potential energy, 18
Potentiometers, electric, 84
Power, 17
    of ten, 7

Power—cont'd
   of two, 9
   ultrasound, 33
Prefixes, scientific, 13-14
Pregnancy
   therapeutic ultrasound in, 162
   ultrasound exposure in, 160
Preprocessing, 113
PRF; *see* Pulse repetition frequency
Printers, laser, 116
Probe(s), 53; see also Transducer, ultrasound, 53
   components of, 54-56
   convex array, 95, 96
   high frequency multi-plane, 98
   specialty ultrasound, 97-98
Processor, film, 117
Program frequency for quality control, 152-153
Properties of waves, 24-30
Prostate, examination of, 98
PRP; *see* Pulse repetition period
Pseudomass, 133
Pulse, ultrasound, characteristics of, 59-61
Pulse average (PA), 37, 38
Pulse duration (PD), 59, 60
Pulse echo ultrasound, 59, 78-87
   amplitude mode in, 79-82
   basic components of, 106
   brightness mode in, 82-87
   motion mode in, 87
   range equation in, 78-79
Pulse repetition frequency (PRF), 59, 60
   and real-time imaging, 90-91
Pulse repetition period (PRP), 59, 60
Pulse repetition rate, 90-91
   and phased array, 97
Pulsed Doppler, 124-127
Pulsed ultrasound, 78
   frequencies of, 56
Pulse-echo technique, 59
PZT; *see* Zirconate-titanate

Q

QA; *see* Quality assurance
QF; *see* Quality factor
Quadrature phase detector, 124
Quality
   image
      dynamic focusing and, 102
      and frame rates, 90
      and lateral resolution, 73
      and number of transducer elements, 89, 90

Quality—cont'd
   image—cont'd
      side lobes and, 101
   of transducer, 56-57
Quality assurance (QA)
   definition of, 143
   for image-recording devices, 117
Quality control, 143-154
   definition of, 143
   program frequency in, 152-153
   routine measurements and observations for, 148-152
   test devices for, 143-148
Quality factor (QF), 56
Quantities
   base and derived, 10, 11
   basic physical, 15-20
      acceleration, 15
      energy, 17-20
      heat, 19-20
      Newton's laws of motion and, 15-16
      power, 17
      velocity, 15
      work, 16
Quarter wavelength matching layer transducer, 54, 56
Quartz in transducer element, 53
Q-value, 56

R

Radiation
   electromagnetic, 24
   and number of transducer elements, 89, 90
Radiation Measurements Incorporated tissue phantoms, 146-148
Radical ion, free, 156
Range, 70
   dynamic, of receiver, 105
Range equation, 78
Range grating, 126
Range resolution in pulsed Doppler, 124
Rarefaction in acoustic wave, 24
Real-time B-mode ultrasonography, 78
   transducers for, and Doppler transducers in one probe, 130
Real-time imagers, types of, 89
Real-time images, 85
   pulsed Doppler incorporated into, 126
Real-time transducers, 88-104
   annular array in, 97
   characteristics of images, 88-92
   electronic, 93-97
   focusing, 98-102

Real-time transducers—cont'd
   mechanical, 92-93
   specialty, 97-98
   steering, 102-103
Receiver, ultrasound, in image processing and display,
      105-108
Recording, image, 116-117
Rectal insertion, probe for, 98
Reflected wave, decibels versus intensity of, 37
Reflection, 45-49
   acoustic impedance in, 47-49
   angle of incidence in, 45-47
   direction of, 79
   example of, 49
   in scattering, 44
   in ultrasonography, 40, 41
   of ultrasound beam, 36
Reflective shadows, 135-137
Refraction artifacts, 44
Refraction in attenuation of ultrasound beam, 43
Refractive edge shadows, 137, 138
Registration in quality control, 149
Reject, 107
Relaxation time in attenuation of ultrasound beam, 43
Renal cyst, enhancement artifact and, 137, 139
Resolution, 69-77
   axial, 70-73
      computing, 73
      in quality control, 150
      SUAR test object and, 146
   contrast, 69
   lateral, 65-66, 73-77
      computing, 76
      in quality control, 150-151
   longitudinal, 70
   memory, 111
   spatial, 69
Resonant frequency, 56
Reverberation, 92
   in image analysis, 133-134, 135
Rib shadows, artifact from, 136, 137
Ring down artifact, 135, 136
RMI tissue phantoms, 146-148
Rods of AIUM test object, 144, 145
Rotating imagers, 92
Rotating multielement transducers, 92-93

S

SA; *see* Spatial average
Safety, clinical, 161-162
Sample volume, 126

Scales, temperature, 19, 20
Scalpel, ultrasonic, in dentistry, 159
Scan conversion, 108-113
Scan converter, 86
   analog, 108, 110
   digital, 110-113
      beam-intensity artifact and, 141-142
Scattering, 44
Scientific notation, review of, 7-8
Scientific prefixes, 13-14
Scission, molecular, 157
Second, definition of, 13
Second law of motion, 15-16, 17
Sector angle in real-time transducer, 91
Sector field of view, 85
Segmental activation, 93-95
Sensitivity
   receiver, 105
   routine evaluation of, 148
   SUAR test object and, 145
   uniformity, and axial resolution test object,
      144-146
Separation of rods in distance calibration, 148-149
Sequential activation, 93, 94
Shadowing artifact, 134-137
Shift, Doppler, 120
Shift frequency, Doppler; *see* Doppler shift frequency
Short fluid path images, 92
SI; *see* Systeme International d'Unites
Side lobes
   in beam-width artifacts, 141
   and image quality, 101
Signal intensity distortion, 141-142
Signal processing, 107-108
Signal to noise ratio for segmental linear array, 95
Significant figures, review of, 5-6
Sine-cosine potentiometers, 84
Single element transducer, 51, 53-56
Sinusoidal wave, locations along, 28, 29
Size, transducer, and lateral resolution, 73-75
Slice thickness
   evaluation of, 151, 152
   focusing and, 99
   and lateral resolution, 75
Sludge in gallbladder as artifact, 141
Smoothing, 108
Snell's law, 43
   in edge shadows, 137
SONAR, development of, 2, 3
Sonic transducers, examples of, 52
Sonoscope, 3

Sound(s)
  decibel levels for, 35
  isotropic emission of, 61
  and ultrasound, comparison of, with electromagnetic
      radiation, 23
Sound waves, 21-24
SP; *see* Spatial peak
Spatial average (SA), 37, 38
Spatial peak (SP), 37, 38
Spatial pulse length (SPL), 61
  axial resolution and, 70, 72
Spatial resolution, 69
Spatial shift in image, 138
Spatial variation of ultrasound beam, 37, 38
Specialty ultrasound probes, 97-98
Specification of intensity, 36-38
Spectral analysis, Doppler, 128-130
Specular reflection, 44, 45, 135
Speed of light, 8
Spherical wave front, 61
SPL; *see* Spatial pulse length
Static modes, 78
  misregistration artifact in, 139
Steering ultrasound beam, 97, 102-103
Stenosis in blood vessels, Doppler analysis of, 128-130
Stethoscope, ultrasonic, 124
Stiffness, 28
Stones, renal and biliary, ultrasonic removal of, 159
Strain of zirconate-titanate, 53
Subsonic frequencies, 1,2
Surface waves, characteristics of, 24
Surgery, ultrasound in, 159
Surgical clip, artifact from, 135
Swept gain, 106
Systeme International d'Unites (SI), 13
Systems of units, 13

T

TA; *see* Temporal average
Television fields, 115
Television frame, 115
Television monitors, 113
Temperature
  elevation of, in tissue, 155-156
  scales of, 19, 20
  of SUAR test object, 144-145
Temporal average (TA), 37, 38
Temporal peak (TP), 37, 38
Temporal variation in ultrasound beam, 37-38
Ten, power of, 7
Test objects, 143, 143-148

Test objects—cont'd
  AIUM, 144-145
  available, 144
  SUAR, 144-146
TGC; *see* Time-gain compensation
TGC slope, 107
Therapeutic ultrasound
  effects of, 159
  in pregnancy, 162
Thermal conduction, 156
Thermal convection, 156
Thermal effects of ultrasound on tissue, 155-156
Thermal energy, 18-19
Thermal process recording, 117
Thickness
  slice
    evaluation of, 151, 152
    focusing and, 99
    and lateral resolution, 75
  of transducer element, 53-54
Third law of motion, 16, 18
Threshold, dose-response, 160-161
Time
  dependance of biological effects on, 161
  required for reflection, 79
  standard of, 13
Time-gain compensation (TGC), 106-108
  evaluation of, 151
Time-gain compensation slope, 107
Time-motion mode, 87
Tissue(s)
  attenuation coefficients in, 40
  axial resolution in, 73
  interface of, with air, reflection at, 47
  length of near field in, 64
  soft relationship of velocity, frequency, and
      wavelength in, 27
Tissue phantoms, RMI, 146-148
Titanic, locating, 2
TM-mode; *see* Time-motion mode
Tolerance values in quality control, 152
TP; *see* Temporal peak
Train whistle, Doppler effect and, 119
Transducer(s), 51-58
  for amplitude mode, 79
  color-flow, 131
  diameter of, relationship of, to near and far field, 65
  high and low Q, 57
  nonfocused, near field lengths and far field
      divergence of, 66
  number of elements in and image quality, 89, 90

Transducer(s)—cont'd
 phased-array, for pulsed Doppler, 127
 piezoelectric effect and, 51-52
 quality factor and, 56-57
 real-time, 88-104
  annular array in, 97
  characteristics of images, 88-92
  electronic, 93-97
  focusing, 98-102
  mechanical, 92-93
  specialty, 97-98
  steering, 102-103
 relationship of velocity, frequency, and wavelength
  in, 27
 rotating multielement, 92-93
 single-element, 51, 53-56
 size of, and lateral resolution, 73-75
 sonic, examples of, 52
Transducer array, 62
Transducer elements
Transition, near field, far field, 63
Transverse resolution, 73
Transverse waves, characteristics of, 21-24
Tumor neovascularity, evaluation of, 131
Two, power of, 9

U

UCH; *see* Ultrasonic cardiography
Ultrasonic cardiography, 87
Ultrasonic diathermy, 159
Ultrasonic physiotherapy, 156
Ultrasonic stethoscope, 124
Ultrasound
 biological effects of, 155-162
  on animals, 158-159
  clinical safety and, 161-162
  dose-response relationships in, 160-161
  on humans, 159-160
  mechanism of action of, 155-157
  on simple structures, 157-158
 continuous wave, 78
  biological effects of, 158
  Doppler, 123-124, 125
  duty factor of, 61
  frequency of, 56
  in physical therapy, 159
 Doppler, 119-132
  aliasing and, 127-128
  color flow, 130-131
  continuous wave, 123-124, 125
  duplex, 130

Ultrasound—cont'd
 Doppler—cont'd
  mathematics in, 119-121
  pulsed, 124-127
  in spectral analysis, 128-130
 frequencies of, 1
 history of, 1-4
 intensity of, 33-36
 interaction of, with matter, 40-50
  attenuation in, 40-42
  attenuation interactions in, 42-44
  diffraction in, 44-45
  interference in, 45
  reflection of, 45-49
  scattering in, 44
  in tissues, 40
 as mechanical wave, 21
 nature of, 1
 power of, 33
 pulsed, 78
  frequencies of, 56
 specification of, 36-38
Uniformity, SUAR test object and, 145-146
Units, systems of, 13

V

Vaginal insertion, probe for, 98
Valleys in acoustic wave, 24
Variable intensity electron beam, 114
Variables
 acoustic, 33
  amplitude of, 35
 in ultrasound equations, 6
VCR; *see* Video cassette recorder
Veins, false portal, as multipath artifacts, 140
Velocity, 15
 description of, 26
 effects of density and compressibility, 27-28
 of ultrasound in various materials, 26
Vertical axis, 10
Video cassette recorder (VCR) for image recording,
  117
Video envelope after smoothing, 108
Video frame, 115
Viscosity in attenuation of ultrasound beam, 43
Visual inspection of scanner, 151-152

W

Water in ultrasound effects, 157-158
Wave(s)
 electromagnetic, 24

Wave—cont'd
  fundamentals of, 21-32
  mechanical, 21-24
  phase of, 28-30
  properties of, 24-30
Wave equation, 26-27
Wave front, 61-62
  plane, 61-63
  spherical, 61
Wavelength, description of, 24-25
Whistle, dog, 1, 2
Width, beam, and lateral resolution, 75, 76
Wild, John, 3-4
Windows, ultrasound, 40

Windows, ultrasound—cont'd
  B-mode, 85
Word length, binary, 110, 111
Work, 16

X

X axis, 10
X-ray interaction in tissues, 40

Y

Y axis, 10

Z

Zirconate-titanate (PZT) in transducer element, 53

3